# ANALYSIS, GEOMETRY, AND PROBABILITY

## LECTURE NOTES

## IN PURE AND APPLIED MATHEMATICS

1. *N. Jacobson,* Exceptional Lie Algebras
2. *L.-Å. Lindahl and F. Poulsen,* Thin Sets in Harmonic Analysis
3. *I. Satake,* Classification Theory of Semi-Simple Algebraic Groups
4. *F. Hirzebruch, W. D. Newmann, and S. S. Koh,* Differentiable Manifolds and Quadratic Forms (out of print)
5. *I. Chavel,* Riemannian Symmetric Spaces of Rank One (out of print)
6. *R. B. Burckel,* Characterization of C(X) Among Its Subalgebras
7. *B. R. McDonald, A. R. Magid, and K. C. Smith,* Ring Theory: Proceedings of the Oklahoma Conference
8. *Y.-T. Siu,* Techniques of Extension on Analytic Objects
9. *S. R. Caradus, W. E. Pfaffenberger, and B. Yood,* Calkin Algebras and Algebras of Operators on Banach Spaces
10. *E. O. Roxin, P.-T. Liu, and R. L. Sternberg,* Differential Games and Control Theory
11. *M. Orzech and C. Small,* The Brauer Group of Commutative Rings
12. *S. Thomeier,* Topology and Its Applications
13. *J. M. Lopez and K. A. Ross,* Sidon Sets
14. *W. W. Comfort and S. Negrepontis,* Continuous Pseudometrics
15. *K. McKennon and J. M. Robertson,* Locally Convex Spaces
16. *M. Carmeli and S. Malin,* Representations of the Rotation and Lorentz Groups: An Introduction
17. *G. B. Seligman,* Rational Methods in Lie Algebras
18. *D. G. de Figueiredo,* Functional Analysis: Proceedings of the Brazilian Mathematical Society Symposium
19. *L. Cesari, R. Kannan, and J. D. Schuur,* Nonlinear Functional Analysis and Differential Equations: Proceedings of the Michigan State University Conference
20. *J. J. Schäffer,* Geometry of Spheres in Normed Spaces
21. *K. Yano and M. Kon,* Anti-Invariant Submanifolds
22. *W. V. Vasconcelos,* The Rings of Dimension Two
23. *R. E. Chandler,* Hausdorff Compactifications
24. *S. P. Franklin and B. V. S. Thomas,* Topology: Proceedings of the Memphis State University Conference
25. *S. K. Jain,* Ring Theory: Proceedings of the Ohio University Conference
26. *B. R. McDonald and R. A. Morris,* Ring Theory II: Proceedings of the Second Oklahoma Conference
27. *R. B. Mura and A. Rhemtulla,* Orderable Groups
28. *J. R. Graef,* Stability of Dynamical Systems: Theory and Applications
29. *H.-C. Wang,* Homogeneous Branch Algebras
30. *E. O. Roxin, P.-T. Liu, and R. L. Sternberg,* Differential Games and Control Theory II
31. *R. D. Porter,* Introduction to Fibre Bundles
32. *M. Altman,* Contractors and Contractor Directions Theory and Applications
33. *J. S. Golan,* Decomposition and Dimension in Module Categories
34. *G. Fairweather,* Finite Element Galerkin Methods for Differential Equations
35. *J. D. Sally,* Numbers of Generators of Ideals in Local Rings
36. *S. S. Miller,* Complex Analysis: Proceedings of the S.U.N.Y. Brockport Conference
37. *R. Gordon,* Representation Theory of Algebras: Proceedings of the Philadelphia Conference
38. *M. Goto and F. D. Grosshans,* Semisimple Lie Algebras
39. *A. I. Arruda, N. C. A. da Costa, and R. Chuaqui,* Mathematical Logic: Proceedings of the First Brazilian Conference

40. *F. Van Oystaeyen,* Ring Theory: Proceedings of the 1977 Antwerp Conference
41. *F. Van Oystaeyen and A. Verschoren,* Reflectors and Localization: Application to Sheaf Theory
42. *M. Satyanarayana,* Positively Ordered Semigroups
43. *D. L. Russell,* Mathematics of Finite-Dimensional Control Systems
44. *P.-T. Liu and E. Roxin,* Differential Games and Control Theory III: Proceedings of the Third Kingston Conference, Part A
45. *A. Geramita and J. Seberry,* Orthogonal Designs: Quadratic Forms and Hadamard Matrices
46. *J. Cigler, V. Losert, and P. Michor,* Banach Modules and Functors on Categories of Banach Spaces
47. *P.-T. Liu and J. G. Sutinen,* Control Theory in Mathematical Economics: Proceedings of the Third Kingston Conference, Part B
48. *C. Byrnes,* Partial Differential Equations and Geometry
49. *G. Klambauer,* Problems and Propositions in Analysis
50. *J. Knopfmacher,* Analytic Arithmetic of Algebraic Function Fields
51. *F. Van Oystaeyen,* Ring Theory: Proceedings of the 1978 Antwerp Conference
52. *B. Kedem,* Binary Time Series
53. *J. Barros-Neto and R. A. Artino,* Hypoelliptic Boundary-Value Problems
54. *R. L. Sternberg, A. J. Kalinowski, and J. S. Papadakis,* Nonlinear Partial Differential Equations in Engineering and Applied Science
55. *B. R. McDonald,* Ring Theory and Algebra III: Proceedings of the Third Oklahoma Conference
56. *J. S. Golan,* Structure Sheaves over a Noncommutative Ring
57. *T. V. Narayana, J. G. Williams, and R. M. Mathsen,* Combinatorics, Representation Theory and Statistical Methods in Groups: YOUNG DAY Proceedings
58. *T. A. Burton,* Modeling and Differential Equations in Biology
59. *K. H. Kim and F. W. Roush,* Introduction to Mathematical Consensus Theory
60. *J. Banas and K. Goebel,* Measures of Noncompactness in Banach Spaces
61. *O. A. Nielson,* Direct Integral Theory
62. *J. E. Smith, G. O. Kenny, and R. N. Ball,* Ordered Groups: Proceedings of the Boise State Conference
63. *J. Cronin,* Mathematics of Cell Electrophysiology
64. *J. W. Brewer,* Power Series Over Commutative Rings
65. *P. K. Kamthan and M. Gupta,* Sequence Spaces and Series
66. *T. G. McLaughlin,* Regressive Sets and the Theory of Isols
67. *T. L. Herdman, S. M. Rankin, III, and H. W. Stech,* Integral and Functional Differential Equations
68. *R. Draper,* Commutative Algebra: Analytic Methods
69. *W. G. McKay and J. Patera,* Tables of Dimensions, Indices, and Branching Rules for Representations of Simple Lie Algebras
70. *R. L. Devaney and Z. H. Nitecki,* Classical Mechanics and Dynamical Systems
71. *J. Van Geel,* Places and Valuations in Noncommutative Ring Theory
72. *C. Faith,* Injective Modules and Injective Quotient Rings
73. *A. Fiacco,* Mathematical Programming with Data Perturbations I
74. *P. Schultz, C. Praeger, and R. Sullivan,* Algebraic Structures and Applications Proceedings of the First Western Australian Conference on Algebra
75. *L. Bican, T. Kepka, and P. Nemec,* Rings, Modules, and Preradicals
76. *D. C. Kay and M. Breen,* Convexity and Related Combinatorial Geometry: Proceedings of the Second University of Oklahoma Conference
77. *P. Fletcher and W. F. Lindgren,* Quasi-Uniform Spaces
78. *C.-C. Yang,* Factorization Theory of Meromorphic Functions
79. *O. Taussky,* Ternary Quadratic Forms and Norms
80. *S. P. Singh and J. H. Burry,* Nonlinear Analysis and Applications
81. *K. B. Hannsgen, T. L. Herdman, H. W. Stech, and R. L. Wheeler,* Volterra and Functional Differential Equations

82. *N. L. Johnson, M. J. Kallaher, and C. T. Long,* Finite Geometries: Proceedings of a Conference in Honor of T. G. Ostrom
83. *G. I. Zapata,* Functional Analysis, Holomorphy, and Approximation Theory
84. *S. Greco and G. Valla,* Commutative Algebra: Proceedings of the Trento Conference
85. *A. V. Fiacco,* Mathematical Programming with Data Perturbations II
86. *J.-B. Hiriart-Urruty, W. Oettli, and J. Stoer,* Optimization: Theory and Algorithms
87. *A. Figa Talamanca and M. A. Picardello,* Harmonic Analysis on Free Groups
88. *M. Harada,* Factor Categories with Applications to Direct Decomposition of Modules
89. *V. I. Istrătescu,* Strict Convexity and Complex Strict Convexity: Theory and Applications
90. *V. Lakshmikantham,* Trends in Theory and Practice of Nonlinear Differential Equations
91. *H. L. Manocha and J. B. Srivastava,* Algebra and Its Applications
92. *D. V. Chudnovsky and G. V. Chudnovsky,* Classical and Quantum Models and Arithmetic Problems
93. *J. W. Longley,* Least Squares Computations Using Orthogonalization Methods
94. *L. P. de Alcantara,* Mathematical Logic and Formal Systems
95. *C. E. Aull,* Rings of Continuous Functions
96. *R. Chuaqui,* Analysis, Geometry, and Probability
97. *L. Fuchs and L. Salce,* Modules Over Valuation Domains

*Other Volumes in Preparation*

# ANALYSIS, GEOMETRY, AND PROBABILITY

## Proceedings of the First Chilean Symposium on Mathematics

edited by

Rolando Chuaqui
Departamento de Matemática
Pontificia Universidad Católica de Chile
Santiago, Chile

MARCEL DEKKER, INC. New York and Basel

Library of Congress Cataloging in Publication Data

Chilean Symposium of Mathematics (1st : 1981 :
Universidad Técnica Federico Santa María)
Analysis, geometry, and probability.

(Lecture notes in pure and applied mathematics ; 96)
Includes index.
1. Mathematical analysis--Congresses. 2. Geometry--Congresses. 3. Probabilities--Congresses. I. Chuaqui, Rolando Basim. II. Title. III. Series.
QA299.6.C45 1981 515 85-1630
ISBN 0-8247-7419-1



MARCEL DEKKER, INC.
270 Madison Avenue, New York, New York 10016

Current Printing (last digit):
10 9 8 7 6 5 4 3 2 1

PRINTED IN THE UNITED STATES OF AMERICA

# Preface

This volume contains versions, sometimes considerably expanded, of invited addresses and communications for the First Chilean Symposium of Mathematics.

The Society of Mathematics of Chile organized a meeting in 1981, which it called the First Chilean Symposium of Mathematics. Certainly, it was not the first meeting that the society had held. Actually, every year since 1965, there had been several meetings in different Chilean cities. This one, in 1981, however, was considered special. In the first place, it was to be a symposium where most Chilean mathematicians would present the results obtained in their research. In the second place, there would be invited Chilean mathematicians residing within the country and abroad. The main criterion for deciding whom to invite from abroad were the interests of the research groups in Chile for the results obtained by those abroad. Thus, the Symposium represented a balance of what had been done by the Chileans residing in Chile and those abroad working in subjects related to what was done in Chile.

Until 1957, Chilean universities and the country in general had little interest in research in mathematics. There was no degree in mathematics proper and, with very few exceptions, professors were not allowed time for research. In spite of this, there were a few heroic figures who kept abreast of modern developments in mathematics and stimulated students in the science. In 1957, the then Rector of the University of Chile, Juan Gómez Millas, organized a center for mathematical studies, where young and older people could devote time for study and research. Also, some German mathematicians were specially hired. In the next decade, the first undergraduate degrees in mathematics were established in different universities, and, with the aid of the universities and several foreign and international agencies, many young students were sent abroad to complete their graduate studies and obtain Ph.D.s or similar degrees. Many of them returned in the 1970s, but some remained abroad pursuing their careers. Graduate studies also began in Chile during that time.

Thus, we may say that mathematical research is young in Chile. In 1980, the Society of Mathematics of Chile thought it was time to see whether all this effort that the country had made had been successful or not. The verdict is in this volume.

The First Chilean Symposium of Mathematics was held December 18-20, 1981, at the Universidad Técnica Federico Santa María in Valparaíso, Chile. The Organizing Committee consisted of Manuel Bustos and Luis Salinas-Carrasco (Universidad Técnica Federico Santa María), Florencio Utreras (Universidad de Chile), and Irene Mikenberg and Rolando Chuaqui (Pontificia Universidad Católica de Chile). Angela Bau of the Catholic University of Chile was the executive secretary. There was also a Program Committee that selected the invited lecturers and helped in the selection of the articles to be published in this volume. It was formed by: Octavio L. Betancourt (New York University), Ricardo Baeza and Jorge Soto-Andrade (Universidad de Chile), Roberto W. Frucht (Universidad Técnica Federico Santa María), and Guido del Pino, Rolando Rebolledo, and Rolando Chuaqui (Pontificia Universidad Católica de Chile).

The main financial support came from the Universidad Técnica Federico Santa María, Pontificia Universidad Católica de Chile, and Conicyt-Chile, with some additional support from the other Chilean universities. We thank these institutions very much, especially the Universidad Federico Santa María that acted as host to the event. Luis Salinas-Carrasco, then Director of its Department of Mathematics, had the main responsibility for the local arrangements of the meeting.

I would like to thank the members of the Program Committee without whose invaluable help I would not have been able to assemble this volume. Special acknowledgments should be given to Marcel Dekker, Inc. for the preparation of the camera ready copy and the inclusion of this book in their Lecture Notes in Pure and Applied Mathematics series.

Rolando Chuaqui

# Contents

# Contributors

LUC BÉLAIR Department of Mathematics, Yale University, New Haven, Connecticut

OCTAVIO L. BETANCOURT* Courant Institute of Mathematical Sciences, New York University, New York, New York

ROLANDO CHUAQUI Departamento de Matemática, Pontificia Universidad Católica de Chile, Santiago, Chile

GUIDO E. DEL PINO Departamento de Probabilidad y Estadística, Pontificia Universidad Católica de Chile, Santiago, Chile

MANUEL ELGUETA Departamento de Matemática, Pontificia Universidad Católica de Chile, Santiago, Chile

CLAUDIO A. FERNÁNDEZ Departamento de Matemática, Pontificia Universidad Católica de Chile, Santiago, Chile

ROBERTO W. FRUCHT Departamento de Matemáticas, Universidad Técnica Federico Santa María, Valparaíso, Chile

REINALDO E. GIUDICI Departamento de Matemática y Ciencia de la Computación, Universidad Simon Bolivar, Caracas, Venezuela

VICTOR GONZÁLEZ-AGUILERA Departamento de Matemáticas, Universidad Técnica Federico Santa María, Valparaíso, Chile

*Current Affiliation: Computer Sciences Department, The City College of New York, New York, New York

SERVET A. MARTINEZ Departamento de Matemáticas y Ciencias de la Computación, Facultad de Ciencias Físicas y Matemáticas, Universidad de Chile, Santiago, Chile

G. MCFADDEN* Courant Institute of Mathematical Sciences, New York University, New York, New York

PETER MILETTA Departamento de Matemáticas y Ciencias de la Computación, Universidad de Santiago de Chile, Santiago, Chile

JORGE MUJICA Instituto de Matemática, Estatística e Ciência da Computação, Universidade Estadual de Campinas, Campinas, São Paulo, Brazil

ROLANDO REBOLLEDO Departamento de Matemática, Pontificia Universidad Católica de Chile, Santiago, Chile

GONZALO E. REYES Departement de Mathematiques et de Statistique, Université de Montreal, Montreal, Quebec, Canada

WILFRED REYES† Instituto de Matemática, Universidade Estadual de Campinas, Campinas, São Paulo, Brazil

LUIS SALINAS-CARRASCO Departamento de Matemáticas, Universidad Técnica Federico Santa María, Valparaíso, Chile

JORGE SOTO-ANDRADE Departamento de Matemáticas, Facultad de Ciencias, Bãsicas y Farmaceuticas, Universidad de Chile, Santiago, Chile

GUNTHER A. UHLMANN Department of Mathematics, Massachusetts Institute of Technology, Cambridge, Massachusetts

---

Current Affiliations:

*Mathematical Analysis Division, National Bureau of Standards, Gaithersburg, Maryland

†Departamento de Matemáticas, Facultad de Ciencias, Bãsicas y Farmacéuticas, Universidad de Chile, Santiago, Chile

# 1
# Convergence of the Method of Harmonic Balance

PETER MILETTA / Departamento de Matemáticas y Ciencias de la Computación,
Universidad de Santiago de Chile, Santiago, Chile

## I. Introduction

The method of harmonic balance is one of the more well known methods for apprximating the periodic solutions of non-linear ordinary differential equations. It is also one of the more theoretically satisfying methods because of its relation to the Rayleigh-Ritz, Galerkin method. Very little seems to be known, however, about the convergence of the approximations to the actual periodics solution

One paper by Borges, Cesari, and Sanchez [1] considers the first approximation to a special class of equations of second order equations and establishes a method to obtain error bounds on the first coefficient. However for a large subclass of this special class - one that includes the undamped forced Duffing equation - the conditions which are imposed seem impossible to fulfill.

In this paper we apply the same ideas as those developed by the above authors to the general case of the $k^{th}$ approximation. These ideas, which take as their starting point the Alternative Method developed first by Cesari in 1963 [2] extend nicely to the general case and we show by example that in the extended form these results can be applied to the Duffing equation.

## II.

We consider the class of second order differential equations of the form

$$x'' + g(t,x) = 0 \tag{1}$$

There are two cases to be distinguished : the autonomous and the non-autonomous.

A. In the non-autonomous case we will assume that g in defined for $|x| \leq A$, $-\infty < t < \infty$, is $T = 2\Pi/w$ periodic and for each x is intergrable on $[0,T]$.Moreover we ssume that g satisfies the symmetry properties

$$g-t,-x) = -g(t,x) \; ; \; g(\tfrac{T}{4} + t,x) = g(\tfrac{T}{4} - t,x).$$

Further we assume that g is uniformaly Lipschitizian with respect to x (and is hence continuous in x) with Lipschitz constant L for $|x| \leq A$, $-\infty < t < \infty$.

B. In the autonomous case we will assume that g(x) is positive in (0,A] and moreover has the form

$$g(x) = \sigma^2 x + \sum_{j=0}^{n} \alpha_{2j+1} x^{2j+1} + g^*(x)$$

where $|g^*(x)| \leq Cx^{2n+3}$ for some constant C and all x, $|x| \leq A$. Moreover it will be assumed that g is Lipschitizian (and hence continuous) with Lipschitz constant L for $|x| \leq A$.

These assumptions placed on g will assure that if a periodic solution exists, it will have the Fourier Series expansion

$$x(t) \sim \sum_{j=0}^{\infty} a_{2j+1} \sin(2j+1)wt$$

## III.

To determine the $k^{th}$ approximation

$$x_k(t) = a_1^k \sin wt + a_3^k \sin 3wt + \ldots a_{2k+1}^k \sin(2k+1)wt \tag{2}$$

to the T periodic solution of (1) by the method of harmonic balance, we substitute (2) in the expression $x'' + g(t,x)$ to obtain

$$-\sum_{j=0}^{k} a_{2j+1}^k (2j+1)^2 w^2 \sin(2j+1)wt + g(t, \sum_{j=0}^{k} a_{2j+1}^k \sin(2j+1)wt) \tag{3}$$

and choose the coefficients $a^k_{2j+1}$ so that the first k harmonics in (3) are zero. This is equivalent to having the numbers $a^k_{2j+1}$ satisfy the system of equations

$$- a_{2j+1}(2j+1)^2w^2 + \frac{2}{T}\int_0^T g(t, \sum_{i=0}^k a_{2j+1}\sin(2i+1)wt)\sin(2j+1)wtdt = 0$$

for j = 0, 1, 2, ..., k. (4)

These equations, following [1], will be called the determining equations for the $a^k_{2j+1}$. For the present we will assume that the system has the solutions

$a^k_1$, $a^k_3$,..., $a^k_{2k+1}$ and further that $\sum_{j=0}^k |a^k_{2j+1}| \leqslant A_0 < A$. Finally we set

$$x_k(t) = \sum_{j=0}^k a^k_{2j+1}\sin(2j+1)wt.$$

In the autonomous case the determining equations will have the form

$$[\sigma^2 - (2j+1)^2w^2]a^k_{2j+1} + \frac{2}{T}\int_0^T \sum_{j=0}^n \alpha_{2j+1} (\sum_{i=0}^k a^k_{2j+1} \sin(2i+1)wt)^i \sin(2j+1)wtdt$$

$$+ \frac{2}{T}\int_0^T g^* (\sum_{i=0}^k a^k_{2j+1} \sin(2i+1)wt)\sin(2j+1)wtdt = 0$$

For $\sum_{i=0}^k |a^k_{2j+1}|$ sufficiently small, the assumption placed on $g^*(x)$ will assure that this system of equations will define $w^2$ as a function of the constants $a^k_1$, $a^k_3$, ..., $a^k_{2k+1}$. If $\sigma \neq 0$, $w^2$ will be positive for small $\sum_{j=0}^k |a^k_{2j+1}|$ and consequently $w(a^k_1, a^k_3,...,a^k_{2k+1})$ is defined and the $k^{th}$ approximation will be

$$x_k(t) = \sum_{j=0}^k a_{2j+1}\sin(2j+1)w(a^k_1, ..., a^k_{2k+1})t$$

If $\sigma = 0$ then the fact that g(x) is assumed positive in (0,A] will assure that the first non zero $\alpha$ will be positive. Again for $\sum_{j=0}^k |a^k_{2j+1}|$ sufficiently the equations will define w as a function of the coefficients $a^k_{2j+1}$. This will be implicitely assumed in the following.

IV.

Let $S_c$ denote the Banach space of all continuous fuctions x(t), $-\infty < t < \infty$ which are $T = \frac{2\Pi}{w}$ periodic and which furthetmore satisfy the symetry conditions $x(t) = -x(-t)$, $x(\frac{T}{4} - t) = x(\frac{T}{4} + t)$ with norm $\|x\| = \max_{0 \leq t \leq T} |x(t)|$. All of the elements of $S_c$ have Fourier series expansions in terms on sine functions :

$$x(t) \sim \sum_{j=0}^{\infty} b_{2j+1} \sin(2j+1)wt$$

where

$$b_{2j+1} = \frac{2}{T} \int_0^T x(t) \sin(2j+1)wt dt$$

We define the operators

$$P^k x = \sum_{j=0}^{k} b_{2j+1} \sin(2j+1)wt$$

and

$$Hx = -\sum_{j=0}^{\infty} \frac{b_{2j+1}}{(2j+1)^2 w^2} \sin(2j+1)wt$$

We have immediately the following

Theorem I : a) For each k, the operator $P^k$ is bounded, linear, and idempotent,

b) The operator H is bounded, $\|H\| \leq \frac{\Pi}{2w^2}$ , and furthermore commutes with $P^k$ for all k.

c) $P^k H(I - P^k) = 0$ for all k.

Proof : The proof in this case is essentially the same as the proof of Theorem I in [ 1] .

The operator $H(I - P^k)$ is obviously linear. To find bounds on $\|H(I - P^k)\|$, we note that if

$$K_k(t,\beta) = \begin{cases} -\beta + \frac{4}{\Pi}(\sum_{j=0}^{k} \frac{\sin(2j+1)wt \sin(2j+1)w\beta}{(2j+1)^2} & 0 \leq \beta \leq t \leq \frac{T}{4} \\ -t + \frac{4}{\Pi}(\sum_{j=0}^{4} \frac{\sin(2j+1)wt \sin(2j+1)w\beta}{(2j+1)^2} & 0 \leq t \leq \beta \leq \frac{T}{4} \end{cases}$$

then

$$H(I - P^k)x + \int_0^{\frac{T}{4}} K_k(t,\beta) x(\beta) d\beta$$

Consquently

$$\|H(I - P^k)\| = \max_{0 \leq t \leq \frac{T}{4}} \int_0^{\frac{T}{4}} |K_k(t,\beta)| \, d\beta .$$

To calculate this integral we make change of variables $wt = \xi$ , $w\beta = \eta$ and obtain

$$\|H(I - P^k)\| = \frac{1}{w^2} \max_{0 \leq \xi \leq \frac{\Pi}{2}} \int_0^{\frac{\Pi}{2}} |\bar{K}_k(\xi,\eta)| \, d\eta \tag{5}$$

where

$$\bar{K}_k(\xi,\eta) = \begin{cases} -\eta + \frac{4}{\Pi}\left(\sum_{j=0}^{k} \frac{\sin(2j+1)\xi \, . \, \sin(2j+1)\eta}{(2j+1)^2}\right) & 0 \leq \eta \leq \xi \leq \frac{\Pi}{2} \\ -\xi + \frac{4}{\Pi}\left(\sum_{j=0}^{k} \frac{\sin(2j+1)\xi \, . \, \sin(2j+1)\eta}{(2j+1)^2}\right) & 0 \leq \xi \leq \eta \leq \frac{\Pi}{2} \end{cases}$$

$$= \begin{cases} \frac{\xi-\eta}{2} - \frac{\xi+\eta}{2} + \frac{2}{\Pi}\sum_{j=0}^{k} \frac{[\cos(2j+1)(\xi-\eta) - \cos(2j+1)(\xi+\eta)]}{(2j+1)^2} & 0 \leq \xi \leq \eta \leq \frac{\Pi}{2} \\ \frac{\eta-\xi}{2} - \frac{\eta+\xi}{2} + \frac{2}{\Pi}\sum_{j=0}^{k} \frac{[\cos(2j+1)(\xi-\eta) - \cos(2j+1)(\xi+\eta)]}{(2j+1)^2} & \leq \eta \leq \xi \leq \frac{\Pi}{2} \end{cases}$$

Now using the fact that functions 1, cosx, cos2x, etc. form a basis for $L^2$ $[0,\Pi]$ we may write

$$\frac{\xi+\eta}{2} = \frac{\Pi}{4} - \frac{2}{\Pi} \sum_{j=0}^{\infty} \frac{\cos(2j+1)(\xi+\eta)}{(2j+1)^2}$$

$$\frac{\xi-\eta}{2} = \frac{\Pi}{4} - \frac{2}{\Pi} \sum_{j=0}^{\infty} \frac{\cos(2j+1)(\xi-\eta)}{(2j+1)^2}$$

etc. Substituting these in the expresion for $\bar{K}_k(\xi,\eta)$ we have

$$\bar{K}_k(\xi,\eta) = \frac{2}{\Pi} \sum_{j=k+1}^{\infty} \frac{\cos(2j+1)(\xi+\eta) - \cos(2j+1)(\xi-\eta)}{(2j+1)^2}$$

$$= -\frac{4}{\Pi} \sum_{j=k+1}^{\infty} \frac{\sin(2j+1)\xi \, \sin(2j+1)\eta}{(2j+1)^2} .$$

Thus

$$\|H(I - P^k)\| = \frac{4}{\Pi w^2} \max_{0 \leq \xi \leq \frac{\Pi}{2}} \int_0^{\frac{\Pi}{2}} \left| \sum_{j=k+1}^{\infty} \frac{\sin(2j+1)\xi \, \sin(2j+1)\eta}{(2j+1)^2} \right| d\eta$$

An upper bound to $\|H(I - P^k)\|$ can be obtained via the Cauchy-Schwartz inequality since

$$\int_0^{\frac{\Pi}{2}} \Big| \sum_{j=k+1}^{\infty} \frac{\sin(2j+1)\,\sin(2j+1)}{(2j+1)^2} \Big|^2 \, d\,\eta = \int_0^{\frac{\Pi}{2}} \sum_{j=k+1}^{\infty} \frac{\sin^2(2j+1)\eta}{(2j+1)^4}$$

$$= \frac{\Pi}{4} \sum_{j=k+1}^{\infty} \frac{1}{(2j+1)^4}$$

Since $\sum_{j=0}^{\infty} \frac{1}{(2j+1)^4} = \frac{\Pi^4}{96}$ we have

$$B_k = \| H(I - P^k) \| \leqslant \frac{\sqrt{2}}{w^2} \left( \frac{\Pi^4}{96} - \sum_{j=0}^{k} \frac{1}{(2j+1)^4} \right)^{\frac{1}{2}} .$$

Thus

$$B_o \leqslant \frac{.17131}{w^2} \;, B_1 \leqslant \frac{.08298}{w^2} \;, B_2 \leqslant \frac{.03826}{w^2} \;, B_3 \leqslant \frac{.025128}{w^2} \;, B_4 \leqslant \frac{.018141}{w^2}$$

The integral (5) of course can be approximated numerically. In each case the maximun value is found to be obtained for $\xi = \frac{\Pi}{2}$ . We find

$$B_o = \frac{.14454}{w^2} \;, B_1 = \frac{.05524}{w^2} \;, B_2 = \frac{.029517}{w^2} \;, B_3 = \frac{.01856}{w^2} \;, B_4 = \frac{.01286}{w^2} .$$

V.

Now let $\overline{S}_c = \{x \mid x \in S_c, \|x\| \leqslant A \}$

and let $gx = g(t,x(t))$. We define the operators

$$F^k x = - H(I - P^k)gx$$

and

$$J_k x = P^k x + F^k x - H(I - P^k)gx$$

mapping $\overline{S}_c$ into $S_c$.

We choose the numbers $c_1^k$ , $c_3^k$ , ..., $c_{2k+1}^k$, $d^k$ such that $c_{2j+1}^k > 0$, $0 < \sum_{j=0}^{k} c_{2j+1}^k < d^k$ and let $b_1^k$ ,...., $b_{2k+1}^k$ be any k + 1 numbers such that $|b_{2j+1}^k - a_{2j+1}^k| \leqslant c_{2j+1}^k$. Define

$$S_c^k(b_1^k, \ldots, b_{2k+1}^k) = \{ x \mid \in S_c, P^k x = \overline{x}_k, \| x - \overline{x}_k \| \leqslant d^k \}$$

where $\overline{x}_k(t) = \sum_{j=0}^{k} b_{2j+1}^k \sin(2j+1)wt$. The space $S_c^k(b_1^k,\ldots,b_{2k+1}^k)$ is nonempty as it contains $\overline{x}_k(t)$. Furthermore being a closed subspace of the complete space $\overline{S}_c$, $S_c^k$ is complete with the norm $\| \;\|$.

Finally let $\gamma^k = \| H(I - P^k)gx_k \|$, $Q^k = \| \sum_{j=0}^{k} a_{2j+1}^k \sin(2j+1)wt \|$.

We observe that if $Q^k + d^k \leqslant A$ then $S_c^k (b_1^k, \ldots, b_{2k+1}^k) \subset \bar{S}_c$ .

Theorem II : Suppose that $B_k L < 1$, $Q^k + d^k \leqslant A$, and moreover that $\sum_{j=0}^{k} c_{2j+1}^k + \gamma^k < (1 - B_k L)d^k$. Then $J_k : S_c^k(b_1^k, \ldots, b_{2k+1}^k) \to S_c(b_1^k, \ldots, b_{2k+1}^k)$ is a contraction and has a fixed point $y(t)$. The fixed point satisfies

$$\| y(t) - \bar{x}_k(t) \| \leqslant d^k - \sum_{j=0}^{k} c_{2j+1}^k .$$

Proof : Let $y \in S_c^k(b_1^k, \ldots, b_{2k+1}^k)$ then

$$\text{i)}\quad P^k(J_k(y)) = P^k(\bar{x}_k - Hgy + HP^k gy) = \bar{x}_k$$

since $P^k H = P^k H P^k$

$$\text{ii)}\quad \text{If } z = P^k y - H(I - P^k)gy \text{ then}$$

$$\begin{aligned} \| z - x_k \| &= \| P^k y - H(I - P^k)gy - x_k \| = \| \bar{x}_k - x_k - H(I - P^k)gy \| \\ &= \| \bar{x}_k - x_k - H(I - P^k)g\,(y - x_k) - H(I - P^k)gx_k \| \\ &\leqslant \sum_{j=0}^{k} c_{2j+1}^k + B_k L d^k + \gamma^k < d^k . \end{aligned}$$

Thus i) and ii) imply that $J_k : S_c(b_1^k, \ldots, b_{2k+1}^k) \to S_c(b_1^k, \ldots, b_{2k+1}^k)$

To see that this is a contraction mapping, let $y_1 = J_k x_1$, $y_2 = J_k x_2$; then

$$\begin{aligned} \| y_1 - y_2 \| &= \| P^k x_1 - P^k x_2 + H(I - P^k)g\,(x_2 - x_1) \| = \| H(I - P^k)g\,(x_2 - x_1) \| \\ &= B_k L \, \| x_2 - x_1 \| . \end{aligned}$$

Finally with $y(t)$ as the fixed point, we observe that

$$\begin{aligned} \| y - \bar{x}_k \| &= \| P^k y - H(I - P^k)gy - \bar{x}_k \| = \| H(I - P^k)gy \| \\ &\leqslant \| H(I - P^k) \| \cdot \| g(y - x_k) \| + \| H(I - P^k)gx_k \| \\ &= B_k L d^k + \gamma^k < d^k - \sum_{j=0}^{k} c_{2j+1}^k \end{aligned}$$

Observation : Since $B_k \to 0$ and $\gamma^k \to 0$ as $k \to \infty$ we may always choose $k$ large enough so as to satisfy the inequality $\sum_{j=0}^{k} c_{2j+1}^k + \gamma^k < (1 - B_k L)d^k$.

Now once we have a fixed point $y(t) \in S_c(b_1^k,\ldots,b_{2k+1}^k)$ we note that $y = P^k y - H(I - P^k)gy$ is twice continuously differentiable and moreover satisfies the differential equation

$$y'' + gy = P^k y'' + P^k gy$$

or more clearly

$$y''(t) + g(t,y(t)) = P^k(y''(t) + g(t,y(t)).$$

At this point the ralation between the ideas of Borges, Cesari, and Sanchez and the Alternative Method of Cesari is most clearly seen. The problem of determining the existence of a periodic solution to (1) is seen to be equivalent to the existence of a function $y(t) \in S_c^k(b_1^k,\ldots,b_{2k+1}^k)$ which satisfies

a) $y = P^k y - H(I - P^k)gy$

and

b) $P^k(y''(t) + g(t,y(t))) = 0$

The existence of a solution to a) is given by the above theorem. The alternative problem then involves showing that we may choose the constants $b_1^k,\ldots,b_{2k+1}^k$ so that the first k + 1 harmonics of $y'' + g(t,y)$ are zero. That is that equation b) is satisfied.

A complete discussion of the Alternative Method may be found in Hale [3].

Thus equation (1) will have y(t) as a T-periodic solutions if we can show that we may choose the constant $b_1^k,\ldots,\ b_{2k+1}^k$ so that the first k + 1 harmonics of $y'' + g(t,y)$ are zero. But since $y = b_1^k \sin wt + \ldots + b_{2k+1}^k \sin(2k+1) + \ldots$ we see that this is equivalent to showing that the system of equations

$$-(2j+1)^2 w^2 b_{2j+1}^k + \frac{2}{T}\int_0^T g(t,y(t))\sin(2j+1)wt\,dt = 0 \qquad (6)$$

for $j = 1,\ldots,k$ has solution $b_1^k,\ldots,b_{2k+1}^k$ hopefully close to $a_1^k,\ldots,a_{2k+1}^k$

To this end we set for $(\alpha_0,\ldots,\alpha_k) \in R^{k+1}$

$$\beta_j(\alpha_0,\ldots,\alpha_k) = \frac{2}{T}\int_0^T g(t,x_\alpha(t))\sin(2j+1)wt\,dt$$

for $j = 0,\ldots,k$, where

$$x_\alpha(t) = \alpha_0 \sin wt + \alpha_1 \sin 3wt + \ldots + \alpha_k \sin(2k+1)wt$$

and

$$\tilde{\beta}_j(\alpha_0 \ldots,\alpha_k) = \frac{2}{T} \int_0^T g(t,y_\alpha(t))\sin(2j+1)wtdt$$

for $j = 1,\ldots, k$ where

$$y_\alpha(t) = \alpha_0 \sin wt + \alpha_1 \sin 3wt + \ldots + \alpha_k \sin(2k+1)wt + \ldots$$

is the fixed point whose existence is given by Theorem II.

We observe that :

$$|\tilde{\beta}_j (\alpha_0,\ldots, \alpha_k) - \beta_j (\alpha_0,\ldots,\alpha_k) | \leqslant$$

$$\frac{2}{T} \int_0^T |g(t,y_\alpha(t)) - g(t,x_\alpha(t))| \, |\sin(2j+1)wt|dt$$

$$\leqslant \frac{\Pi}{4} L(B_k L d^k + \gamma_\alpha^k)$$

where

$$\gamma_\alpha^k = \| H(I - P^k)gx_\alpha \| .$$

We consider the mapping of $R^{k+1} \to R^{k+1}$ given in terms of the $i^{th}$ coordinates by

$$(u(\alpha_0, \ldots, \alpha_k))_i = - (2i+1)^2 w^2 \alpha_i + \beta_i(\alpha_0, \ldots, \alpha_k)$$

and

$$(U(\alpha_0,\ldots, \alpha_k))_i = - (2i+1)^2 w^2 \alpha_i + \tilde{\beta} (\alpha_0,\ldots, \alpha_k)$$

We can think of the mapping U as a perturbation of the mapping u in the sense that $U = u + (U - u)$ where

$$\| (U - u)_i \| \leqslant \frac{\Pi}{4} L(B_k L d^k + \gamma_\alpha^k ).$$

Since $(u(a_1^k , a_3^k ,\ldots,a_{2k+1}^k))_i = 0$ for $i = 0,1,\ldots,k$ , if the Jacobian of the transformation u is non-singular at the point $(a_1^k,\ldots,a_{2k+1}^k)$ then u will map some neighborhood of $(a_1^k,\ldots,a_{2k+1}^k)$ onto some neighborhood of 0 in $R^{k+1}$.

We consider the closed neigborhood

$$N = \{(a_1^k + d_1,\ldots,a_{2k+1}^k + d_{2k+1})| \; |d_{2j+1}| \leqslant c_{2j+1}^k, \; j = 0,1,\ldots,k\}$$

about the point $(a_1^k,\ldots,a_{2k+1}^k)$ determined by the constants $c_1^k,\ldots,c_{2k+1}^k$.

If it is possible to choose these constants so that u(N) contains the neighborhood

$$D = \{x_0,\ldots,x_k) \mid |x_j| \leq \frac{\Pi}{4} L(b_k Ld^k + \gamma^k)$$

then U(N) will contain the origen. That is, the system of equations (6) will have a solution in the neighborhood N. Thus :

Theorem III : Given the above assumtions, if $U(N) \supset D$ then the system of equations (6) has a solution $(b_1^k,\ldots,b_{2k+1}^k)$ which satisfies $|b_{2j+1}^k - a_{2j+1}^k| \leq c_{2j+1}^k$ for $j = 0, \ldots, k$.

Corollary : Under the above assumtions and those of Theorem III the fixed point y(t)

$$y(t) = b_1 \sin wt + b_3 \sin 3wt + \ldots + b_{2k+1}\sin(2k+1)wt + \ldots$$

is a T-periodic solution and moreover for $j = 0, 1, \ldots, k$

$$|b_{2j+1} - a_{2j+1}| \leq c_{2k+1},\ \|y(t) - x_k(t)\| \leq d^k, \text{ and}$$

$$\|y(t) - x_k(t)\| \leq B_k Ld^k + \gamma^k \text{ where}$$

$$x_k(t) = a_1^k \sin wt + a_3^k \sin 3wt + \ldots + a_{2k+1}^k \sin(2k+1)wt$$

is the $k^{th}$ approximation given by the method of Harmonic Balance.

VI.

Example : We consider the equation

$$x'' + x - \frac{1x^3}{6} = \frac{-1}{16}\sin t$$

The Lipschitz constant be taken equal to 1 for $|x| \leq 2$. The first approximation $x = a_0^1 \sin t$ leads to the equation

$$\frac{-1}{8}(a_0^1)^3 = \frac{-1}{16}$$

for the constant $a_0^1$. Immediately $a_0^1 = .7937005$, and $x_0(t) = .7937005\sin t$. Approximately $\gamma^0 = .0023148$ and $1 - B_0L = .85546$. The inequalities which the numbers $c_1^0$ and $d^0$ must satisfy are

a) $c_1^0 + .0023148 \leq .85546d^0$

b) $u_0(a_1^0 - c_1^0) = \frac{-1}{8}(a_1^0 - c_1^0)^3 + \frac{1}{16} \leq \frac{-4}{\Pi}(.14454d^0 + .0023148)$

c) $u_0(a_1^0 + c_1^0) = \frac{-1}{8}(a_1^0 + c_1^0)^3 + \frac{1}{16} \geq \frac{4}{\Pi}(.14454d^0 + .0023148)$

Taking $.85546d^o = c_1^o + .0023148$ these are equivalent to finding the first positive $c_1^o$ for which

$$.2362352c_1^o - .2976377(c_1^o)^2 + .125(c_1^o)^3 + \frac{4}{\pi}(.14454c_1^o + .0026494) < 0$$

and

$$.2362352c_1^o + .2976377(c_1^o)^2 + .125(c_1^o)^3 - \frac{4}{\pi}(.14454c_1^o + .0026494) > 0$$

Unfortunately the root of the second equation

$$-.125(c_1^o)^3 + .2976377(c_1^o)^2 + .0520844c_1^o - .0033733 = 0$$

is larger than the value of $a_o^1$.

Thus we consider the second approximation

$$x_1(t) = a_1^2 \sin t + a_3^2 \sin 3t$$

The equations for $a_1^2$ and $a_3^2$ are

$$\frac{-1}{6}(3(a_1^2)^3 - 3(a_1^2)^2(a_3) + (a_1^2)(a_3^2)^2) = \frac{-1}{16}$$

and

$$8(a_3^2) - \frac{1}{6}\left(\frac{1}{4}(a_3^2)^3 + \frac{3}{4}(a_1^2)^2(a_3^2) - \frac{1}{4}(a_1^2)^3\right) = 0$$

The roots of this system are

$$a_1^2 = .794021 \ ; \quad a_3^2 = .004128$$

Moreover by direct calculation we have $B_1 = .05524$ , $\gamma_o^3 = .000534$

$$\begin{aligned} u_1(a_1^2 + d_1^2, a_3^2 + d_3^2) = &- .3406922(d_1^2) - .29725(d_1^2)^2 - .125(d_1^2)^3 \\ &+ 0777458(d_1^2) - .0006667(d_3^2)^2 + .197833(d_1^2)(d_3^2) \\ &+ .125(d_1^2)^2(d_3^2) + .166(d_1^2)(d_3^2)^2, \end{aligned}$$

$$\begin{aligned} u_2(a_1^2 + d_1^2, a_3^2 + d_3^2) = &-.7.9212115(d_3^2) - .000002(d_3^2)^2 - .416667(d_3^2)^3 \\ &-.4740975(d_1^2) - .5955(d_1^2)^2 - .25(d_1^2)^3 \\ &-.1985(d_1^2)(d_3^2) - .0005(d_3^2) - .125(d_3^2)^3. \end{aligned}$$

The inequality which $d^2$ , $c_1^2$, and $c_3^2$ must satisfy is

$$.94476d^2 \geqslant .000534 + c_1^2 + c_3^2$$

and with $d^2 = .0005652 + 1.05847(c_1^2 + c_3^2)$ we have $\frac{4}{\pi}(\gamma^1 + LB_1 d^2) =$ $.0012848 + .0584699(c_1^2 + c_3^2)$. By direct calculation we find that for

$|d_1| \leqslant .0283938$ and $|d_3| \leqslant .0025641$ we have

$$u_1(a_1^2 + .0283938,\ a_3^2 + d_3) - \frac{4}{\Pi}(\gamma^1 + LB_1d^2) > 0$$

$$u_1(a_1^2 + -.0283938,\ a_3^2 + d_3) + \frac{4}{\Pi}(\gamma^1 + LB_1d^2) < 0$$

$$u(a_1^2 + d_1,\ a_3^2 - .0025641) - \frac{4}{\Pi}(\gamma^1 + LB_1d^2) > 0$$

$$u(a_1^2 + d_1,\ a_3^2 + .0025641) + \frac{4}{\Pi}(\gamma^1 + LB_1d^2) < 0$$

Thus we may take $c_1^2 = .028394$ , $c_3^2 = .002564$; and thus $d^2 = .0341043$. Consequently the equation will have a periodic solution of period $2\pi$ $y(t) \sim b_1\sin t + b_3\sin 3t$ where $|b_1 - .794021| < .028394$, $|b_3 - .004128| < .002564$ and such that $\| y(t) - .794021\sin t - .004128\sin 3t \| \leqslant .0341043$.

As a final remark we note that the usual form of the undamped forced Duffing equation

$$x'' + \omega_0^2 x + \alpha x^3 = k\cos\Omega t$$

can be treated by this method by first making the change of variables $\tau = t + \frac{\Pi}{2\Omega}$.

References

1) Borges, C.A., L. Cesari, and D.A. Sanchez (1974), Functional Analysis and the Method of Harmonic Balance, Q. Appl. Math. 32, 475 - 464, 61

2) L. Cesari. " Funcional Analysis and Periodic Solutions of Nonlinear Differential Equations ", Contributions to Differential Equations I, Wiley, N.Y., 1963. pp. 149 - 187.

3) J. Hale, Ordinary Differential Equations, Wiley - Interscience, New York, 1969.

# 2
# Resolvent Estimates in Some Exterior Regions

CLAUDIO A. FERNÁNDEZ / Departamento de Matemática, Pontificia Universidad Católica de Chile, Santiago, Chile

## 1. Introduction

Throughout this paper, $\Omega$ denotes an exterior region in $\mathbb{R}^3$, that is, an open connected set such that its complement, the obstacle $O = \mathbb{R}^3 - \Omega$, is compact.

We consider the system consisting of a quantum mechanical particle moving freely in $\Omega$. With an adequate choice of units, the Hamiltonian H for such system is the selfadjoint realization of the negative Laplace operator in $\Omega$ with Dirichlet boundary conditions on $\partial\Omega$. The operator H acts on the Hilbert space $H = L^2(\Omega)$. The unit vectors of $H$ will be called the states of the system.

We denote by $H_0$ the free Hamiltonian $-\Delta_{\mathbb{R}^3}$, viewed as a selfadjoint operator on $H_0 = L^2(\mathbb{R}^3)$.

---

This work is part of the author's Ph.D. thesis at the University of Rochester, under the direction of Professor R. Lavine, N November, 1981. The author's research is partially supported by grant # 17/82 of the Dirección de Investigacion de la Pontificia Universidad Católica de Chile.

It is physically clear and mathematically known (see for example [5] and [9]) that there is a good scattering theory for the pair $H_0$, $H$. This fact can be proven by using the theory of smooth operators. The main ingredient in this proof is certain resolvent estimate which we first show using a standard non constructive approach. This result and a generalization of it are then applied to study the sojourn time of the different states.

In Section 3, we present a method of computing explicit resolvent estimates in the case when the obstacle is star-shaped. We then obtain explicit upper bounds for the sojourn time and for the spectral concentration. Finally, in Section 4 we apply the previous results to conclude the nonexistence of resonances in the star-shaped case.

It is worth noting that resonances do occur when the obstacle has a partially opened cavity ([2]). It can be shown that there are states initially concentrated in the cavity which spend an unusually large amount of time there. Here, explicit lower bounds for the sojourn time are needed. These results, together with estimates on the location of the resonant energies, will be presented elsewhere.

## 2. A non-constructive approach

The characteristic function of a set $S \subset \mathbb{R}^3$ will be denoted by $\chi_S$. If $f$ is a function on a set $S$, then we also denote by $f$ the operator "multiplication by f", defined in $L^2(S)$.

In what follows, $r_0 > 0$ is large enough so that the obstacle $\mathcal{O}$ is contained in the ball $B(0,r_0)$.

For $r \geq r_0$, we write $\Omega(r) = \Omega \cap B(0,r)$ and $\chi_r = \chi_{\Omega(r)}$.

Given $r_1 \leq r_2$, $A(r_1,r_2)$ denotes the shell $\{x : r_1 \leq |x| \leq r_2\}$.

If $U \subset \mathbb{R}^3$ is an open set, we set $\| \ \|_{0,U} = \| \ \|_{L^2(U)}$ and we simply write $\| \ \| = \| \ \|_{0,\Omega}$.

Given a nonnegative integer m, $H^m(U)$ denotes the Sobolev space consisting of all functions in $L^2(U)$ whose first m derivatives (in the distributional sense) also belong to $L^2(U)$. The norm in $H^m(U)$ is given by

$$\|\phi\|^2_{m,U} = \sum_{|\alpha| \leq m} \left\| \frac{\partial^\alpha \phi}{\partial x^\alpha} \right\|^2_{0,U},$$

for all $\phi \in H^m(U)$. Here, we have used the standard multi-index notation.

LEMMA 1. *Assume that the region* $\Omega$ *satisfies the segment property* (see [1]). *Then, given* $r \geq r_0$ *and a bounded interval* $I \subset [0,\infty]$, *there exists a positive constant* c *such that*

$$\|\chi_r (H-\lambda-i\varepsilon)^{-1} \chi_r\| \leq c ,$$

*uniformly for* $\varepsilon \neq 0$ *and* $\lambda \in I$.

The constant c depends only on r, I, and the obstacle $\mathcal{O}$.

PROOF. We prove the stronger result that $\|\chi_{5r}(H - \lambda - i\varepsilon)^{-1}\chi_r\|$ is uniformly bounded for $\varepsilon \neq 0$ and $\lambda \in I$.

Let us suppose that above is false. Then, for each nonnegative integer n, there exist $\varepsilon_n \neq 0, \lambda_n \in I$ and $\phi_n \in L^2(\Omega)$ such that

(i) $\|\phi_n\| = 1$ and

(ii) $c_n^2 = \|\chi_{5r}(H - \lambda_n - i\varepsilon_n)^{-1} \chi_r\phi_n\|^2 \geq n.$

Since $n \leq \|(H - \lambda_n - i\varepsilon_n)^{-1}\|^2 = \varepsilon_n^{-2}$, the sequence $\{\varepsilon_n\}$ converges to zero. Also, by taking subsequences if necessary, we can assume that $\lim_{n\to\infty} \lambda_n = \lambda$, where $\lambda \in I$.

We write $t_n = \lambda_n + i\varepsilon_n$, $\theta_n = \frac{1}{c_n} \chi_r\phi_n$ and $\psi_n = (H - t_n)^{-1}\theta_n$.

Then $\psi_n \in \mathcal{D}(H)$, $H\psi_n = t_n\psi_n + \theta_n$ and $\|\psi_n\|_{0,\Omega(5r)} = 1$.

We first claim that the sequence $\{\psi_n\}$ is bounded in $H^1(\Omega(4r))$.

Let $\alpha \in C_0^\infty(\mathbb{R}^3)$ be such that $0 \leq \alpha \leq 1$ and

$$\alpha(x) = \alpha(|x|) = \begin{cases} 1 & \text{if} \quad |x| \leq 4r \\ 0 & \text{is} \quad |x| \geq 5r. \end{cases}$$

Then,

$$\int_{\Omega(4r)} |\nabla\psi_n|^2 \leq \int_{\Omega(5r)} \alpha \, \nabla\overline{\psi}_n \cdot \nabla\psi_n$$

$$= -\mathrm{Re} \int_{\Omega(5r)} \nabla\alpha \cdot \nabla\psi_n\overline{\psi}_n - \mathrm{Re} \int_{\Omega(5r)} \alpha \, \Delta\psi_n\overline{\psi}_n$$

$$= \frac{1}{2} \int_{\Omega(5r)} \Delta\alpha|\psi_n|^2 + \mathrm{Re} \int_{\Omega(5r)} \alpha(t_n|\psi_n|^2 + \theta_n\overline{\psi}_n)$$

$$\leq \frac{1}{2} \|\Delta\alpha\|_\infty \|\psi_n\|^2 + \lambda_n\|\psi_n\|^2 + \frac{1}{c_n} \|\phi_n\| \, \|\chi_r\psi_n\|$$

$$\leq \frac{1}{2} \|\Delta\alpha\|_\infty + \lambda_n + \frac{1}{c_n} ,$$

which is bounded indepently of n, thus proving our claim.

Therefore, $\{\psi_n\}$ has a subsequence which is weakly convergent to some $\psi$ in $H^1(\Omega(4r))$. We can denote this subsequence again by $\{\psi_n\}$, by deleting some terms, if necessary. Since $\Omega(4r)$ satisfies the segment property, the inclusion $H^1(\Omega(4r)) \to H^0(\Omega(4r))$ is compact (see [1]); hence, we conclude that the sequence $\{\psi_n\}$ converges strongly to $\psi$ in $H^0(\Omega(4r))$.

We now claim that $\psi$ is a solution of the boundary value problem

$$-\Delta\psi = \lambda\psi \quad \text{in} \quad \Omega(4r),$$

$$\psi = 0 \quad \text{on} \quad \partial\Omega.$$

If $\theta \in C_0^\infty\,(\Omega(4r))$ then

$$\int_{\Omega(4r)} (-\Delta\psi)\theta = \int_{\Omega(4r)} \nabla\psi \cdot \nabla\theta$$

$$= \lim_{n\to\infty} \int_{\Omega(4r)} \nabla\psi_n \cdot \nabla\theta$$

$$= \lim_{n\to\infty} \int_{\Omega(4r)} (-\Delta\psi_n)\theta$$

$$= \lim_{n\to\infty} \int_{\Omega(4r)} (t_n\psi_n + \theta_n)\theta$$

$$= \lambda \int_{\Omega(4r)} \psi\theta ,$$

since $\{\theta_n\}$ converges to zero strongly.

For the boundary condition, let us consider the trace operator T: $H^1(\Omega(4r)) \to H^0(\partial\Omega)$, which is a continuous mapping ([3]).

Given $v \in H^0(\partial\Omega)$ we have

$$\langle T\psi, v\rangle_{0,\partial\Omega} = \langle \psi, T^*v\rangle_{1,\Omega(4r)}$$

$$= \lim_{n\to\infty} \langle \psi_n, T^*v\rangle_{1,\Omega(4r)}$$

$$= \lim_{n\to\infty} \langle T\psi_n, v\rangle_{0,\partial\Omega}$$

$$= 0.$$

Therefore, $\psi$ is a weak solution of above boundary value problem and by elliptic theory for the Laplacian, it is a strong solution.

We want to extend $\psi$ to all $\Omega$, the extension still being an eigenfunction of $-\Delta$ with eigenvalue $\lambda$. For that purpose, let us consider the Green function $e(x,y,t)$, that is, the kernel of the free resolvent $(H_0 - t)^{-1}$ (in our case, in three dimensions, $e(x,t) = (4\pi|x|)^{-1}\exp(-i\sqrt{t}|x|)$, for $\mathrm{Im}\sqrt{t} > 0$, and $e(x,y,t) = e(x-y,t)$).

We denote by $S_r$ the sphere $\{x : |x| = r\}$ and by $\nu$ the outwards normal.

Let $|x| > 2r$. Then, Green's Theorem implies that

$$\int_{S_{2r}} (\psi_n(y)\, \partial_\nu\, e(x,y,t_n) - \partial_\nu \psi_n(y)\, e(x,y,t_n))d\sigma(y)$$

$$= \int_{|y|>2r} (\Delta\psi_n(y)e(x,y,t_n) - \psi_n(y)\, \Delta e(x,y,t_n))dy$$

$$= \int_{|y|>2r} (-t_n\psi_n(y)\, e(x,y,t_n) - \frac{1}{c_n}\, \chi_r(y)\theta_n(y)e(x,y,t_n) - \psi_n(y)\, \Delta e(x,y,t_n))dy$$

$$= \int_{|y|>2r} \psi_n(y)\, (-\Delta-t_n)e(x,y,t_n)dy.$$

By the fundamental property of the Green function we therefore obtain.

$$\psi_n(x) = \int_{|y|=2r} (\psi_n(y)\partial_\nu\, e(x,y,t_n) - \partial_\nu\, \psi_n(y)e(x,y,t_n))d\sigma(y), \tag{1}$$

for $|x| > 2r$.

But, as $n \to \infty$, $\psi_n \to \psi$ strongly in $H^0(\Omega(4r))$, thus $-\Delta\psi_n = t_n\psi_n + \theta_n \to \lambda\psi = -\Delta\psi$ strongly in $H^0(\Omega(4r))$. Therefore, by ellipticity of the Laplacian, $\psi_n \to \psi$ strongly in $H^2(A(r,3r))$. (Here we have used that if $U$ and $V$ are open sets in $\mathbb{R}^3$ satisfying $\overline{U} \subset V$, then there exists a positive constant $c$ such that

$$\|u\|_{2,U} \leq c(\|\Delta u\|_{0,V} + \|u\|_{0,V}), \text{ for all } u \in H^2(V)).$$

Hence, $\psi_n \to \psi$ strongly in $H^1(S_{2r})$ and we can take the limit under the integral sign in (1). We obtain that as $n \to \infty$,

$$\psi_n(x) \to \int_{S_{2r}} (\psi(y)\partial_\nu e(x,y,\lambda) - \partial_\nu\psi(y)e(x,y,\lambda))d\sigma(y), \qquad (2)$$

uniformly for x in any compact set contained in $\{x : |x| > 2r\}$.

But, by taking limits as n approaches infinity to both sides of (1), we conclude

$$\psi(x) = \int_{S_{2r}} (\psi(y)\partial_\nu e(x,y,\lambda) - \partial_\nu\psi(y)e(x,y,\lambda))d\sigma(y), \qquad (3)$$

for all $x \in A(2r,4r)$.

We then extend $\psi$ to all of $\Omega$ by using the same formula (3).

It follows that $\psi \in H^1(\Omega)$ is a solution of the boundary value problem

$$-\Delta\psi = \lambda\psi \quad \text{in} \quad \Omega,$$
$$\psi = 0 \quad \text{on} \quad \partial\Omega.$$

Moreover, $\psi$ also satisfies a radiation condition at infinity, because it is a superposition of outgoing fundamental solutions. Hence, since $\lambda$ is a nonnegative number, $\psi$ must vanish identically (by a classical theorem of F. Rellich, $\psi = 0$ in a neighborhood of infinity and, as a solution of above elliptic equation, $\psi$ is analytic in the connected region $\Omega$ and therefore $\psi = 0$ in all of $\Omega$ (see [13]).

But this leads to a contradiction: $\{\psi_n\}$ converges uniformly to zero in compact sets in $\{x : |x| > 2r\}$, hence strongly in $H^0(\Omega(5r))$ and $\|\psi_n\|_{0,\Omega(5r)} = 1$, for all n. #

We now use Lemma 1 to show local H-smoothness of certain operators. First we review the definition ([4], [6]).

DEFINITION 1. _A closed operator A acting on a Hilbert space H is said to be smooth with respect to a selfadjoint operator_

*H on $H$ (or, simply H-smooth) if $\mathcal{D}(H) \subset \mathcal{D}(A)$ and there exists a positive constant c such that*

$$2|\varepsilon| \quad \| A(H-\lambda-i\varepsilon)^{-1}\phi\|^2 \leq c^2\|\phi\|^2,$$

*for all $\phi \in H$, $\varepsilon \neq 0$ and $\lambda \in \mathbb{R}$. The smallest such c is called the H-norm of A and it is denoted by $\|A\|_H$.*

DEFINITION 2. *We say that A is H-smooth in a Borel set $I \subset \mathbb{R}$ if $AE_H(I)$ is H-smooth.*

Here, $E_H(\cdot)$ denotes the projection valued spectral measure associated with the selfadjoint operator H.

Next result gives a (neccesary and) sufficient condition for local H-smoothness. For a proof we mention [6].

THEOREM 1. (Lavine). *Let I be a Borel set in $\mathbb{R}$. Suppose that there exists a positive constant c such that*

$$2|\varepsilon| \quad \| A(H - \lambda- i\,\varepsilon)^{-1}\phi\|^2 \leq c\|\phi\|^2,$$

*uniformly for $\phi \in H$, $\varepsilon \neq 0$ and $\lambda \in I$. Then, A is H-smooth in the closure of I.*

COROLLARY 1. *Lef f be a bounded measurable function with bounded support contained in $\Omega$. Then, there exists a positive constant $c = c(f,I,\mathcal{O})$ such that*

$$|\varepsilon| \quad \| f(H-\lambda-i\varepsilon)^{-1}\|^2 \leq c \ ,$$

*uniformly for $\varepsilon \neq 0$ and $\lambda \in I$.*

In other words, f is H-smooth in any bounded interval I.

*Proof*. Let $r \geq r_0$ such that supp $f \subset B(0,r)$ and let $M = \sup \{|f(x)| : x \in \Omega\}$. Then

$$|\varepsilon| \; \| f(H-\lambda-i\varepsilon)^{-1}\|^2 \le M^2|\varepsilon| \; \|\chi_r(H-\lambda-i\varepsilon)^{-1}\|^2$$

$$= M^2|\varepsilon| \; \|\chi_r(H-\lambda-i\varepsilon)^{-1} \; (H-\lambda+i\varepsilon)^{-1}\chi_r\|$$

$$= M^2 \, \tfrac{1}{2} \; \|\chi_r((H-\lambda-i\varepsilon)^{-1} \; -(H-\lambda+i\varepsilon)^{-1})\chi_r\|$$

$$\le M^2 \sup_{\varepsilon\neq 0} \; \|\chi_r(H-\lambda-i\varepsilon)^{-1}\chi_r\| .$$

The Corollary then follows from Lemma 1. #

The corresponding result for $H_0$ can be proven by direct computation [10], since the kernel of the free resolvent is explicitly known. The following stronger result also holds for $H_0$.

PROPOSITION 1. _Let_ L _be a first order differential operator acting on_ $H = L^2(\Omega)$. _Suppose that_ L _has smooth coefficients with compact support in_ $\Omega$. _Then_ L _is_ H-_smooth in any bounded interval_ $I \subset \mathbb{R}$.

_Proof._ Let K be the support of L and let U be an open neighborhood of K. Then, by elliptic estimates,

$$\| L(H-\lambda-i\varepsilon)^{-1}\varphi\|^2 = \| L(H-\lambda-i\varepsilon)^{-1}\varphi\|^2_{0,K}$$

$$\le c(\| (H-\lambda)(H-\lambda-i\varepsilon)^{-1}\varphi\|^2_{0,U} + \| (H-\lambda-i\varepsilon)^{-1}\varphi\|^2_{0,U})$$

$$\le c\|\varphi\|^2_{0,U} + c \; \| (H-\lambda-i\varepsilon)^{-1}\varphi\|^2_{0,U}$$

$$\le c\|\varphi\|^2 + c\|\chi_U(H-\lambda-i\varepsilon)^{-1}\varphi\|^2,$$

for all $\varphi \in H$.

Hence, $|\varepsilon| \; \| L(H-\lambda-i\varepsilon)^{-1}\|^2 \le c|\varepsilon| + c|\varepsilon| \; \|\chi_U(H-\lambda-i\varepsilon)^{-1}\|^2$ and the Propositition follows from Lemma 1. #

We now introduce more notation to be used in the rest of this section.

Let $j \in C_0^\infty(\mathbb{R}^3)$ be a cut-off function satisfying $0 \leq j \leq 1$ and

$$j(x) = j(|x|) = \begin{cases} 0 & \text{if} \quad |x| \leq r_0 \\ 1 & \text{if} \quad |x| \geq 2r_0. \end{cases}$$

We also denote by j the bounded operator given by "multiplication by j(x)" and by $L_j$ differential operator $-\Delta j - 2\nabla j \cdot \nabla$, both operating in any adequate Hilbert space.

The Rollnik class $R$ is the set of all functions in $\mathbb{R}^3$ satisfying

$$\iint \frac{|f(x)||f(y)|}{|x-y|} \, dx \, dy < \infty .$$

It is known ([12]) that $L^{3/2}(\mathbb{R}^3) \subset R$ and that for any $f^2 \in R$, $\| |f| (H_0 - z)^{-1} |f| \|$ is uniformly bounded in z. In particular any such $|f|$ is $H_0$-smooth.

THEOREM 2. Given a bounded function $f^2 \in R$, set $g = f/\Omega$ (the restriction of f to $\Omega$). Then g is H-smooth in any bounded interval $I \subset \mathbb{R}$.

Proof. Write $z = \lambda + i\varepsilon$, $R_0 = R_0(z) = (H_0 - z)^{-1}$ and $R = R(z) = (H-z)^{-1}$.

Consider $\psi \in H$. Then $\phi = R\psi \in \mathcal{D}(H)$, $j\phi \in \mathcal{D}(H_0)$ and $(H_0 - z) j\phi = j(H-z)\phi + L_j \phi$.

Hence, $jR\psi = R_0 j\psi + R_0 L_j R\psi$.

We therefore obtain the resolvent equation:

$$jR = R_0 j + R_0 L_j R.$$

Thus

$$\begin{aligned} \sqrt{|\varepsilon|} \, \|gR\| &\leq \sqrt{|\varepsilon|} \, \|g(1-j)R\| + \sqrt{|\varepsilon|} \, \|gR_0 j\| + \sqrt{|\varepsilon|} \, \|gR_0 L_j R\| \\ &\leq \sqrt{|\varepsilon|} \, \|g(1-j)R\| + \sqrt{|\varepsilon|} \, \|gR_0\| + \sqrt{|\varepsilon|} \, \|gR_0 L_j R\| . \end{aligned}$$

The first term is bounded uniformly for $\varepsilon \neq 0$ and $\lambda \in I$, since $g(1-j)$ has bounded support in $\Omega$.

The second term is bounded independently of $\lambda$ and $\varepsilon$, because $g^2$ is in the Rollink class.

To estimate the third term, define $\tilde{g}$ by

$$\tilde{g} = \begin{cases} g & \text{if } x \notin A(r_0, 2r_0) \\ \max\{g,1\} & \text{if } x \in A(r_0, 2r_0). \end{cases}$$

Then $\tilde{g}^2 \in R$ and there exists $c_1 > 0$ such that $\|\tilde{g}R_0\tilde{g}\| \leq c_1$, uniformly in z.

Therefore, the third term is bounded by

$$\sqrt{|\varepsilon|}\ \|gR_0\ \chi_{A(r_0,2r_0)}\|\ \|L_jR\| \leq \sqrt{|\varepsilon|}\ \|L_jR\|\ \|\tilde{g}R_0\tilde{g}\|$$

$$\leq c_1\ \sqrt{|\varepsilon|}\ \|L_jR\|\text{, which is}$$

uniformly bounded for $\varepsilon \neq 0$ and $\lambda \in I$, by Propositition 1. #

Theorem 2 can be used to obtain that the operator $H = -\Delta_\Omega$ is absolutely continuous and, morever, that it is unitarily equivalent to $H_0$ ([2]). From the viewpoint of scattering theory this means that any state with Hamiltonian H must be scattered towards infinity. In other words, a quantum mechanical particle moving freely in the exterior region $\Omega$ is expected to spend only a finite amount of time in any bounded neighborhood of the obstacle $O = \mathbb{R}^3 - \Omega$. The resolvent estimate of Theorem 2 provides a proof of this fact, whenever the particle has energy restricted to a bounded set.

The projection onto the one dimensional space spanned by a vector $\phi \in H$ is denoted by $P_\phi$, so that $P_\phi(\psi) = \langle\phi,\psi\rangle\phi$, for any $\psi \in H$.

Following [7], we define the <u>time of sojourn of a unit vector</u> $\phi \in H$ to be

$$\tau_H(\phi) = \int_{-\infty}^{\infty} \| P_\phi e^{-iHt}\phi \|^2 dt$$

$$= \int_{-\infty}^{\infty} |\langle \phi, e^{-iHt}\phi \rangle|^2 dt.$$

The quantity $\| P_\phi e^{-iHt}\phi \|^2$ is the probability that, at time t, the system is still in its initial state $\phi$. Thus, $\tau_H(\phi)$ is the expected amount of time the system will spend in its original state.

A related quantity is the expected _amount of time_ (_transit time_) the particle with initial wave function $\phi$ spends in a region $R \subset \Omega$. We denote it by $t_H(\phi, R)$ so that

$$t_H(\phi, R) = \int_{-\infty}^{\infty} \| \chi_R \, e^{-iHt}\phi \|^2 dt$$

Throughout the rest of this section, I denotes a bounded interval in $\mathbb{R}^+$.

Next result provides time independent estimates for the sojourn and transit times; it can be obtained from the theory of smooth operators.

THEOREM 3. _Let A be a bounded operator on_ $\mathcal{H}$. _Suppose that_ $\phi$ _is a unit vector in_ $\mathcal{H}$ _which satisfies_ $E_H(I)\phi = \phi$. _Then_,

$$2|\varepsilon| \| A(H-\lambda-i\varepsilon)^{-1}\phi \|^2 \le \int_{-\infty}^{\infty} \| Ae^{-iHt}\phi \|^2 dt \le \overline{\lim_{\varepsilon \to 0}} \sup_{\lambda \in I} 2|\varepsilon| \, \| A(H-\lambda-i\varepsilon)^{-1} \|^2 ,$$

_for all_ $\lambda \in I$ _and_ $\varepsilon \in \mathbb{R}$ _with_ $\varepsilon \neq 0$.

_Proof_. See [7].

THEOREM 4. _Let_ $\phi \in \mathcal{H}$ _be a unit vector satisfying_ $E_H(I)\phi = \phi$.

a) _Let_ $r \ge r_0$. _Then, there exists a positive constant_ $c_1 = c_1(r, \mathcal{O}, I)$ _such that_ $t_H(\phi, \Omega(r)) \le c_1$, _and_

b) <u>If, furthermore, there exists</u> $f \in L^3(\Omega)$ <u>such that</u> $f^{-1}\phi \in H$ <u>then, there exists a positive constant</u> $c_2 = c_2(f,\mathcal{O},I)$ <u>such that</u> $\tau_H(\phi) \leq c_2\|f^{-1}\phi\|^2$.

<u>Proof.</u> The assertion a) follows immediately from Corollary 1 and Theorem 3.

In order to prove b), we note that again by Theorem 3,

$$\tau_H(\phi) \leq \overline{\lim_{\varepsilon\to 0}} \sup_{\lambda\in I} \|P_\phi(H-\lambda-i\varepsilon)^{-1}\|^2.$$

But, for any $\psi \in H$, we have that

$$\|P_\phi(H-\lambda-i\varepsilon)^{-1}\psi\|^2 = |\langle\phi,(H-\lambda-i\varepsilon)^{-1}\psi\rangle|^2$$

$$\leq \|f^{-1}\phi\|^2 \; \|f(H-\lambda-i\varepsilon)^{-1}\psi\|^2.$$

Hence,

$$\tau_H(\phi) \leq \overline{\lim_{\varepsilon\to 0}} \sup_{\lambda\in I} 2|\varepsilon| \; \|f(H-\lambda-i\varepsilon)^{-1}\|^2 \; \|f^{-1}\phi\|^2.$$

Finally, since $f^2 \in L^{3/2}(\Omega)$, we have that $f^2$ belongs to the Rollnik class $R$. It follows from Theorem 2 that the operator given by multiplication by f is H-smooth in the interval I, thus proving part b) of the Theorem.

The dependence $c_2 = c_2(f,\mathcal{O},I)$ can be derived from the proof of Theorem 2.

#

It is easy to see that a unit vector $\phi \in H$ will satisfy the condition in part b) of above Theorem if $r^{1+\varepsilon}\phi$ is square integrable in a neighborhood of infinity.

Thus, when the initial state $\phi$ has energy localized in a bounded interval, both the sojourn time of $\phi$ and its transit time trough a bounded region are finite, at least whenever $\phi$ decays sufficiently quickly.

<u>Remark.</u>

The proof of Lemma 1 appears essentially in [13] (with the Neuman boundary condition). The result is used there to obtain the so called limiting absorption principle and it is based on a classical result of F. Rellich [11] on the growth of solutions of the reduced wave equation in exterior regions. The proof also uses the selection theorem of F. Rellich which is proven in [1] for domains satisfying the "segment property". For other boundary restrictions under which the selection theorem still holds see [13].

## 3. Star-shaped obstacles

In this section we deal with the case in which the obstacle $\mathcal{O}$ is star-shaped, that is, there exists $x_0 \in \mathcal{O}$ such that for all $x \in \mathcal{O}$, the line segment joining $x$ and $x_0$ is contained in $\mathcal{O}$.

One may expect that a star-shaped obstacle cannot trap a quantum mechanical particle for a long time. In fact, the maximum expected time that an interacting state takes to leave a bounded subset of $\Omega$ is not too large compared to the time that a free state (with Hamiltonian $H_0 = -\Delta_{\mathbb{R}^3}$) would spend in the same region. This can be seen from explicit resolvent estimates, which we shall obtain in the star-shaped case.

Again, $H$ denotes the selfadjoint realization of the negative Laplacian in $\Omega = \mathbb{R}^3 - \mathcal{O}$ with Dirichlet boundary conditions on $\partial\Omega$ and operating on $\mathcal{H} = L^2(\Omega)$.

For $x \in \Omega$, we write $r = |x|$ and $\phi_r = \nabla\phi \cdot \frac{x}{r}$, where $\phi$ is any differentiable function in $\Omega$.

The outwards normal on $\partial\Omega$ will be denoted by $\nu$.

LEMMA 2. <u>Let $g(r)$ be a bounded function in $\Omega$ such that $g'$ is measurable and satisfies $2g \geq rg' \geq 0$ in $\Omega$. Suppose that $\phi \in C^2(\Omega), \phi = 0$ on $\partial\Omega$ and $\lambda \geq 0$. Then,</u>

$$\lambda \int_\Omega g'|\phi|^2 \leq 2\mathrm{Re} \int_\Omega \frac{g}{r} (r\bar{\phi})_r(-\Delta-\lambda)\phi. \tag{1}$$

Furthermore, if g" is also measurable in $\Omega$, then

$$\int_{\Omega} (\lambda g' - g''h - g'h' - g'h^2)|\phi|^2 \leq 2\mathrm{Re} \int_{\Omega} \frac{g}{r} (r\bar{\phi})_r(-\Delta-\lambda)\phi, \tag{2}$$

where h(r) and h'(r) are measurable functions in $\Omega$.

Proof. The following identity is straightforward:

$$2\mathrm{Re}\, \frac{g}{r} (r\bar{\phi})_r(-\Delta-\lambda)\phi = \nabla \cdot (-2\mathrm{Re}\, \frac{g}{r} (\bar{\phi} + r\bar{\phi}_r)\nabla\phi + g|\nabla\phi|^2 \frac{x}{r}$$

$$-g(\lambda + \frac{1}{r^2})|\phi|^2 \frac{x}{r}) + \frac{g'}{r^2} |(r\phi)_r|^2 + \frac{2g-rg'}{r} (|\nabla\phi|^2) - |\phi_r|^2) + \lambda g'|\phi|^2$$

By integrating this equality in $\Omega$ and using that $\phi$ vanishes on $\partial\Omega$ we obtain that

$$- \int_{\partial\Omega} (-2\mathrm{Re} g\, \bar{\phi}_r\bar{\phi}_\nu + g|\nabla\phi|^2 \frac{x}{r} \cdot \nu)d\sigma + \int_{\Omega} \frac{2g-g'r}{r} (|\nabla\phi|^2 - |\phi_r|^2)$$

$$+ \int_{\Omega} \frac{g'}{r^2} |(r\phi)_r|^2 + \lambda \int_{\Omega} g'|\phi|^2 = 2\mathrm{Re} \int_{\Omega} \frac{g}{r} (r\bar{\phi})_r(-\Delta-\lambda)\phi.$$

But, on $\partial\Omega$ we have that $\nabla\phi = \phi_\nu \nu$. Hence, the integral on the boundary reduces to

$$\int_{\partial\Omega} g|\phi_\nu|^2 \frac{x}{r} \cdot \nu \, d\sigma .$$

Since the obstacle is star-shaped with respect to the origin, $\frac{x}{r} \cdot \nu \geq 0$ on $\partial\Omega$. Hence, by eliminating some non-negative terms, we obtain

$$\lambda \int_{\Omega} g'|\phi|^2 + \int_{\Omega} \frac{g'}{r^2} |(r\phi)_r|^2 \leq 2\mathrm{Re} \int_{\Omega} \frac{g}{r} (r\bar{\phi})_r(-\Delta-\lambda)\phi, \tag{3}$$

and (1) follows since $g' \geq 0$ in $\Omega$.

Now, assuming that g" is measurable in $\Omega$, we have that

$$0 \leq \int_\Omega \frac{g'}{r^2} \, |(r\phi)_r - hr\phi|^2$$

$$= \int_\Omega \frac{g'}{r^2} \, |(r\phi)_r|^2 + \int_\Omega g'h^2|\phi|^2 - 2\mathrm{Re} \int_\Omega \frac{g'h}{r^2} \, r\bar{\phi}(r\phi)_r$$

$$= \int_\Omega \frac{g'}{r^2} \, |(r\phi)_r|^2 + \int_\Omega g'h^2|\phi|^2 + \int_\Omega \nabla \cdot \left(\frac{g'h}{r^2} \, \frac{x}{r}\right) \, |r\phi|^2$$

$$= \int_\Omega \frac{g'}{r^2} \, |(r\phi)_r|^2 + \int_\Omega (g'h^2 + g''h + g'h')|\phi|^2 \, ,$$

where $h(r)$ and $h'(r)$ are measurable functions in $\Omega$.

We then obtain

$$\int_\Omega \frac{g'}{r^2} \, |(r\phi)_r|^2 \geq - \int_\Omega (g''h + g'h' + g'h^2)|\phi|^2 \, ,$$

and the second inequality also follows from (3). #

PROPOSITION 2. <u>Let</u> $R \geq r_0$. <u>Then</u>,

$$\lambda|\varepsilon| \, \|\chi_R(H-\lambda-i\varepsilon)^{-1}\|^2 \leq 2 + 2R\sqrt{\lambda+|\varepsilon|} \, ,$$

<u>for all</u> $\lambda \geq 0$ <u>and</u> $\varepsilon \neq 0$.

<u>Proof</u>. Let $\mathcal{D}$ be the set of all $\phi \in C^2(\Omega)$ with $\phi = 0$ on $\partial\Omega$.

We choose $g = g(r)$ by

$$g(r) = \begin{cases} r & \text{if } r \leq R \\ R & \text{is } r \geq R \end{cases}$$

Then, $g$ satisfies the conditions for inequality (1) of Lemma 2 and therefore,

$$\lambda \int_\Omega \chi_R|\phi|^2 \leq 2\mathrm{Re} \int_\Omega \frac{g}{r} \, (r\bar{\phi})_r(-\Delta-\lambda)\phi$$

$$= 2\mathrm{Re} \int_\Omega \left(\frac{g}{r} \, \bar{\phi} + g\bar{\phi}_r\right)(-\Delta-\lambda)\phi \leq 2(\|\phi\| + R\|\phi_r\|) \, \|(H-\lambda)\phi\| \, ,$$

for all $\phi \in \mathcal{D}$.

Given $\phi \in \mathcal{D}$, we set $\psi = (H-\lambda-i\varepsilon)\phi$ . Then, it follows that

$$\lambda\|\chi_R\phi\|^2 \leq 2(\|(H-\lambda-i\varepsilon)^{-1}\| \ \|\psi\| + R\|\phi_r\|)\ \|(H-\lambda)(H-\lambda-i\varepsilon)^{-1}\|\,\|\psi\|$$

$$\leq 2(\frac{1}{|\varepsilon|}\|\psi\| + R\|\phi_r\|)\ \|\psi\|$$

But,

$$\|\phi_r\|^2 \leq \sum_{i=1}^{3} \langle D_i\phi,\ D_i\phi\rangle$$

$$= \langle\phi,\ H\phi\rangle \leq \|\phi\|\ \|(H-\lambda)\phi\| + \lambda\|\phi\|^2 \leq (\frac{1}{|\varepsilon|} + \frac{\lambda}{|\varepsilon|^2})\ \|\psi\|^2$$

Therefore,

$$\lambda\|\chi_R(H-\lambda-i\varepsilon)^{-1}\psi\|^2 \leq 2(\frac{1}{|\varepsilon|} + \frac{R\sqrt{\lambda+|\varepsilon|}}{|\varepsilon|})\ \|\psi\|^2$$

for all $\psi \in (H-\lambda-i\varepsilon)\mathcal{D}$.

To complete the proof we only need to show that $(H-\lambda-i\varepsilon)\mathcal{D}$ is dense in $\mathcal{H}$. Let $v \in C_0^\infty(\Omega)$. Then, by regularity of elliptic equations, the solution u of $(-\Delta-\lambda-i\varepsilon)u = v$, in $\Omega$ with $u = 0$, on $\partial\Omega$, is in $C^\infty(\Omega)$; in particular $u \in \mathcal{D}$ and we conclude that $(H-\lambda-i\varepsilon)\mathcal{D}$ contains $C_0^\infty(\Omega)$, which is dense in $\mathcal{H}$.

The resolvent estimate given by Proposition 2 breaks down if $\lambda$ is close to zero. However, by a different choice of the function g(r) in Lemma 2 (2), we obtain:

PROPOSITION 3. *Consider* $R \geq r_0$. *Then,*

$$|\varepsilon|\|\chi_R(H-\lambda-i\varepsilon)^{-1}\|^2 \leq 2\cdot(\lambda + \frac{1}{4R^2})^{-1}(1 + 2R\sqrt{\lambda+|\varepsilon|}),$$

*for all* $\lambda \geq 0$ *and* $\varepsilon \neq 0$.

*Proof*. We first choose $g = g(r)$ in $\Omega$ to be

$$g(r) = \begin{cases} r & ,\ \text{if } r \leq R \\ 2R - \dfrac{R^2}{r} & ,\ \text{if } r \geq R\,. \end{cases}$$

Hence, we have that g" is measurable, $2g \geq rg' \geq 0$, $g \leq 2R$ and $\frac{g}{r} \leq 1$.

Then, by considering $h(r) = \frac{1}{2r}$, it follows from Lemma 2 (2),

$$\int_{\Omega(R)} (\lambda + \frac{1}{4r^2})|\phi|^2 + \int_{r \geq R} \frac{R^2}{r^2}(\lambda + \frac{5}{4r^2})|\phi|^2$$

$$\leq 2 \text{ Re} \int_{\Omega} \frac{g}{r} (r\bar{\phi})_r(-\Delta-\lambda)\phi.$$

Therefore,

$$(\lambda + \frac{1}{4R^2}) \|\chi_R\phi\|^2 \leq \langle(\lambda + \frac{1}{4r^2})\phi,\phi\rangle$$

$$\leq 2 \text{ Re} \langle\frac{g}{r}\phi + g\phi_r, (H-\lambda)\phi\rangle$$

$$\leq 2(\|\phi\| + 2R\|\phi_r\|) \|(H-\lambda)\phi\|.$$

The proof proceeds as in the proof of Proposition 2.

#

THEOREM 5. *Let $\phi \in \mathcal{D}(H)$ be a unit vector such that $E_H([a,b])\phi = \phi$, where $[a,b]$ is an interval in $[0,\infty)$. Assume that $R \geq r_0$. Then,*

a) $t_H(\phi,\Omega(R)) \leq \dfrac{4(1 + 2R\sqrt{a})}{a + 1/4R^2}$, *and*

b) *If, furthermore, a is positive then,*

$$t_H(\phi,\Omega(R)) \leq \frac{4}{a} + \frac{4R}{\sqrt{a}}.$$

*Proof.* By Theorem 3, we have that

$$t_H(\phi,\Omega(R)) \leq \overline{\lim_{\varepsilon\to 0}} \sup_{\lambda\in[a,b]} 2|\varepsilon| \|\chi_R(H-\lambda-i\varepsilon)^{-1}\|^2$$

Hence, from Proposition 3, we conclude

$$t_H(\phi,\Omega(R)) \leq \lim_{\varepsilon\to 0} \sup_{\lambda\in[a,b]} 4(\lambda + 1/4R^2)^{-1}(1 + 2R\sqrt{\lambda+|\varepsilon|})$$

$$= 4(a + 1/4R^2)^{-1}(1 + 2R\sqrt{a}),$$

which proves part a).

Assertion b) follows from Theorem 3 and Proposition 2. #

Throuhgout the rest of this section $f(r)$ denotes the function in $\Omega$ defined by $f(r) = \frac{1}{r_1}$, if $r \leq r_1$ and $f(r) = 1/r$, if $r \geq r_1$, where $r_1 > 0$ is a fixed radius.

The next result generalizes Propositions 2 and 3; it will be used to obtain explicit estimates for the sojourn time of certain states in $H$.

PROPOSITION 4. Consider $\lambda \geq 0$ and $\varepsilon \in \mathbf{R} - \{0\}$. Then,

$$|\varepsilon| \, \| F(H-\lambda-i\varepsilon)^{-1} \|^2 \leq 2\left(\frac{1}{r_1^2} + \frac{2\sqrt{\lambda+|\varepsilon|}}{r_1}\right) ,$$

where $F = F(r)$ is a function defined in $\Omega$ which satisfies

$$F^2(r) = \begin{cases} \lambda f^2 + \frac{1}{4} f^4 & , \text{ if } r \leq r_1 \\ \lambda f^2 + \frac{5}{4} f^4 & , \text{ if } r \geq r_1 . \end{cases}$$

Proof. We consider the function $g(r) = r/r_1^2$, if $r \leq r_1$ and $g(r) = 2/r_1 - 1/r$, if $r \geq r_1$, defined in $\Omega$.

We also set $h(r) = \frac{1}{2r}$, in $\Omega$. Then it follows that

$$\lambda g' - g''h - g'h' - g'h^2 = \lambda g' - \frac{g''}{2r} + \frac{g'}{4r^2} \geq F^2.$$

Therefore, by Lemma 2(2), we have.

$$\| F\varphi \|^2 \leq 2 \left( \| \tfrac{g}{r} \|_\infty \, \|\varphi\| + \|g\|_\infty \, \|\varphi_r\| \right) \| (H-\lambda)\varphi \|$$

$$= 2\left(\frac{1}{r_1^2} \|\varphi\| + \frac{2}{r_1} \|\varphi_r\|\right) \| (H-\lambda)\varphi \| ,$$

for all $\varphi \in \mathcal{D}(H)$.

Finally, given a unit vector $\psi \in H$, consider $\psi = (H-\lambda-i\varepsilon)^{-1}\varphi$ and, as in Proposition 2, we obtain

$$|\varepsilon| \| F(H-\lambda-i\varepsilon)^{-1}\psi\|^2 \leq 2(\frac{1}{r_1^2} + \frac{2\sqrt{\lambda+|\varepsilon|}}{r_1}) \|\psi\|^2$$

which concludes the proof.

THEOREM 6. <u>Let $\phi \in H$ be a unit vector such that</u> $f^{-1}\phi \in H$. <u>Assume furthermore that</u> $E_H([a,b])\phi = \phi$, <u>where</u> $[a,b]$ <u>is an interval contained in</u> $\mathbb{R}^+$. <u>Then, the sojourn time of</u> $\phi$ <u>satisfies</u>

$$\tau_H(\phi) \leq \frac{4}{r_1\sqrt{a}} (2 + \frac{1}{r_1\sqrt{a}}) \|f^{-1}\phi\|^2$$

<u>Proof.</u> By Theorem 3 we have that

$$\tau_H(\phi) \leq \overline{\lim_{\varepsilon\to 0}} \sup_{\lambda\in[a,b]} 2|\varepsilon| \| P_\phi(H-\lambda-i\varepsilon)^{-1}\|^2 .$$

But, for $\psi \in H$, we obtain

$$\| P_\phi(H-\lambda-i\varepsilon)^{-1}\psi\|^2 = |<\phi,(H-\lambda-i\varepsilon)^{-1}\psi>|^2$$

$$\leq \|f^{-1}\phi\| \; \|f(H-\lambda-i\varepsilon)^{-1}\|^2 \; \|\psi\|^2 .$$

It follows that

$$\tau_H(\phi) \leq \overline{\lim_{\varepsilon\to 0}} \sup_{\lambda\in[a,b]} 2|\varepsilon| \| f(H-\lambda-i\varepsilon)^{-1}\|^2 \; \|f^{-1}\phi\|^2 .$$

Finally, let $F = F(r)$ be as in Proposition 4. Then,

$$\varepsilon\| f(H-\lambda-i\varepsilon)^{-1}\|^2 \leq |\varepsilon| \| \frac{F}{\sqrt{\lambda}} (H-\lambda-i\varepsilon)^{-1}\|^2$$

$$= \frac{2}{\lambda} (\frac{1}{r_1^2} + \frac{2}{r_1} \sqrt{\lambda+|\varepsilon|}).$$

We therefore conclude that

$$\tau_H(\phi) \leq \frac{4}{a} (\frac{1}{r_1^2} + \frac{2\sqrt{a}}{r_1}) \|f^{-1}\phi\|^2 .$$

#

## 4. <u>Resonances</u>

The operator $H = -\Delta_\Omega$, where $\Omega$ is an arbitrary exterior region, has only a continuous spectrum. However, there exist states $\phi$ occupied by the system for a long time, which may be so large that $\phi$ could be mistaken by an observer as a bound state (eigenvector). As shown in [7], the existence of a unit vector $\phi$ satisfying this property is especially noteworthy when $\phi$ is concentrated in position and energy (see also [8]).

As in [7], the _energy width of a unit vector_ $\phi \in H$ _about a real number_ $\lambda_0$ is defined to be

$$\Delta E(\phi,\lambda_0) = \inf\{\varepsilon > 0 : \varepsilon^2 \|(H-\lambda-i\varepsilon)^{-1}\phi\|^2 \geq \tfrac{1}{2}\}$$

By the Spectral Theorem for selfadjoint operators,

$$\varepsilon^2\|(H-\lambda_0-i\varepsilon)^{-1}\phi\|^2 = \varepsilon^2<\phi,((H-\lambda_0)^2 + \varepsilon^2)^{-1}\phi>$$

$$= \int_{\sigma(H)} \frac{\varepsilon^2}{(\lambda-\lambda_0)^2 + \varepsilon^2}\, d<\phi,E_H(\lambda)\phi>,$$

where $E_H(\cdot)$ denotes the projection-valued spectral measure associated with the operator H.

Hence, $\Delta E(\phi,\lambda_0)$ is a small number if and only if the measure $d<\phi,E_H(\lambda)\phi>$ is concentrated near $\lambda_0$. Moreover, $\Delta E(\phi,\lambda_0) = 0$ if and only if $\phi$ is an eigenvector of H with eigenvalue $\lambda_0$.

A lower bound for the sojourn time can be derived from an energy-time uncetainty principle. Physically, this principle expresses that the smaller $\Delta E(\phi,\lambda)$ is, the more stable the state $\phi$ and the longer its lifetime.

THEOREM 7. _If_ $\phi \in H$ _is a unit vector and_ $\lambda \in \mathbb{R}$ _then_,

$$(\tau_H(\phi))^{-1} \leq 2\Delta E(\phi,\lambda) \leq 2\|(H-\lambda)\phi\|$$

_Proof_. See [7].

#

A unit vector $\phi \in \mathcal{D}(H)$ is a _resonant state_ for the exterior problem if:

i) $\phi$ is concentrated in position in a bounded region $R \subset \Omega$, i.e.,

$$\|\phi\|^2_{\Omega-R} = \int_{\Omega-R} |\phi|^2 \text{ is small, and}$$

ii) $\phi$ is also concentrated in energy about a real number $\lambda_0$, i.e.,

$\Delta E(\phi,\lambda_0)$ is small.

Because of Theorem 7, if the system starts out in the state $\phi$, if spends a long time there. Morever, if $\phi$ is supported in R then the transit time through R is large, for

$$\tau_H(\phi) = \int_{-\infty}^{\infty} \|P_\phi e^{-iHt}\phi\|^2 dt \leq \int_{-\infty}^{\infty} \|\chi_R e^{-iHt}\phi\|^2 dt = t_H(\phi,R).$$

Above quantities are compared with the corresponding quantities in the free situation, when considering the Hamiltonian $H_0 = -\Delta$ acting on $\mathcal{H}_0 = L^2(\mathbb{R}^3)$. By our choice of units, the velocity fo a free particle with energy $\lambda_0$ is $v_0 = 2\sqrt{\lambda_0}$, therefore, such particle can be expected to stay in R an amount of time at most $\frac{d}{2\sqrt{\lambda_0}}$, where d is the diameter of R.

Hence, the magnitude of the quotient $\frac{d\Delta E(\phi,\lambda_0)}{2\sqrt{\lambda_0}}$ measures the strength of the resonance. We note that because $\hbar$ was set equal to 1, above quotient is dimensionless, otherwise, its smallness would depend on the units we use.

Finally, we again assume that the obstacle is star-shaped. We want to show that resonances do not occur in this case.

The upper bound for $\tau_H(\phi)$ given in Theorem 6 is valid whenever the state $\phi$ has energy bounded away from zero and satisfies that $r\phi$ is square integrable in a neighborhood of infinity.

On the other hand, let $\varphi \in H$ be a unit vector such that $r^2\phi$ is square integrable at infinity and let a be a positive number. Then,

$$|\varepsilon| \, \|(H-\lambda-i\varepsilon)^{-1}\phi\|^2 \leq |\varepsilon| \, \|F(H-\lambda-i\varepsilon)^{-1}\phi\|^2 \, \|\tfrac{\phi}{F}\|^2$$

$$\leq \frac{2}{r_1} \left(\frac{1}{r_1} + 2\sqrt{\lambda+|\varepsilon|}\right) \|\tfrac{\phi}{F}\|^2 ,$$

where F is as in Proposition 4.

Since $F^2 \geq (\lambda r_1^2 + \frac{1}{4}) f^2$, we have that

$$|\varepsilon| \, \|(H-\lambda-i\varepsilon)^{-1}\phi\|^2 \leq \frac{2}{r_1^4} \cdot \frac{1 + 2r_1\sqrt{\lambda+|\varepsilon|}}{\lambda + \frac{1}{4r_1^2}} \|\frac{\phi}{f^2}\|^2$$

$$\leq \frac{12}{r_1^2} \|\frac{\phi}{f^2}\|^2 ,$$

if $\varepsilon$ is sufficiently small.

THEOREM 8. *Let $\phi \in H$ be a unit vector such that $r^2\phi$ is square integrable at infinity. Then,*

$$\tau_H(\phi) \leq \frac{24}{r_1^2} \|\frac{\phi}{f^2}\|^2 ,$$

*where $r_1$ and f are as in* Proposition 4.

*Proof*. By above computations, we have

$$\tau_H(\varphi) \leq \varlimsup_{\varepsilon\to 0} \sup_{\lambda\geq 0} \; 2|\varepsilon| \, \|(H-\lambda-i\varepsilon)^{-1}\phi\|^2$$

$$\leq \frac{24}{r_1^2} \|\frac{\phi}{f^2}\|^2$$

#

A state which is concetrated in position in a bounded region will certainly satisfy the conditions in Theorem 8. Therefore, its sojourn time will not be too large and, by the energy-time uncertainty principle such state can not constitute a genuine resonance.

## References

[1] S.Agmon, Elliptic boundary value problems, Van Nostrand, 1965.

[2] C.Fernández, Resonances in Obstacle Scattering, Ph.D. Thesis, Univ. of Rochester, 1981.

[3] G.Folland, Introduction to Partial Differential Equations Math. Notes, Princeton University Press, 1976.

[4] T.Kato, Smooth operators and commutators, Studia Mathematica, T.31 (1968).

[5] T.Kato, Scattering theory with two Hilbert spaces, J. of Functional Analysis I (1967), 342-369.

[6] R.Lavine, Commutators and Scattering theory II, a class off one-body problems, Indiana Univ. Math. J., Vol.21 (1972), 643-656.

[7] R.Lavine, Spectral density and sojourn times, in: Atomic Scattering Theory, J. Nutall ed. London, Ontario, Univ. of Western Ontario Press, 1978.

[8] R.Lavine, The classical limit of the number of quantum states, to appear in Classical, Semiclassical and Quantum Mechanical Problems in Mathematics, Chemistry and Physics, Plenum, New York.

[9] M.Reed and B.Simon, Methods of Modern Mathematical Physics III, Scattering Theory, New York, Academic Press, 1980.

[10] M.Reed and B.Simon, Methods of Modern Mathematical Physics, IV, Analysis of Operators, New York, Academic Press, 1978.

[11] F.Rellich, Ein Satz über mittlere konvergenz, Gött. Nachr. 30-35 (1930).

[12] B.Simon, Quantum Mechanics for Hamiltonians defines as Quadratic forms, Princeton series in Physics, Princeton University Press, 1971.

[13] C.Wilcox, Scattering theory for the d'Alambert equation in exterior domains, Lecture Notes in Math. 442, Springer Verlag, 1975.

# 3
# Holomorphic Retractions from Convex Domains to Plane Cross Sections

MANUEL ELGUETA / Departamento de Matemática, Pontificia Universidad Católica de Chile, Santiago, Chile

## §1 Introduction

Let $D$ be a strictly convex domain in $\mathbb{C}^n$ with $C^\infty$ boundary. This means $D = \{z \in D' \mid \rho(z) < 0\}$ where $\rho$ is a $C^\infty$ real valued function defined on a neighborhood $D'$ of $\overline{D}$ and the real Hessian of $\rho$ is strictly positive definite everywhere on $D'$. We will assume that $O \in D$ and that $D$ is bounded. We make the identification $\mathbb{C}^k = \mathbb{C}^k \times \{0\} \times \cdots \times \{0\} \subset \mathbb{C}^n$ and we define $M = D \cap \mathbb{C}^k$.

The purpose of this note is to study the following problem:

Under what conditions does there exist a holomorphic retraction $R : D \to M$ which is $C^\infty$ up to the boundary.

We denote by $H(D)$ the class of holomorphic functions on $D$ and by $C^\infty(\overline{D})$ the class of $C^\infty$ functions on $\overline{D}$. Moreover we put $A^\infty(D) = H(D) \cap C^\infty(\overline{D})$ and we observe that if such a retraction $R : D \to M$ exists it is very easy to obtain a linear multiplicative extension operator $L : A^\infty(M) \to A^\infty(D)$ by defining for $f \in A^\infty(M)$; $Lf(z) = f \circ R(z)$.

Existence and construction of holomorphic retractions have also been studied in a different context and with a different spirit in [1], [2] and [3].

## §2 Statement of the results and some examples

Again we let $D = \{z \in D' \mid \rho(z) < 0\}$ be a strictly convex domain in $\mathbb{C}^n$ and $M = D \cap \mathbb{C}^k$ and we introduce the following notation

$$\varphi_i(\xi) = \frac{\partial\rho}{\partial z_i}(\xi,0) \quad \forall\xi \in \partial M \; ; \quad i = 1,\cdots,n$$

$$\varphi(\xi) = (\varphi_1(\xi),\cdots,\varphi_k(\xi)) \quad \forall\xi \in \partial M$$

$$\operatorname{grad} \rho(z) = \left(\frac{\partial\rho}{\partial z_1}(z),\cdots,\frac{\partial\rho}{\partial z_n}(z)\right) \quad \forall z \in D' \quad .$$

It is well known that the convexity of $D$ and $M$ implies $\operatorname{grad} \rho(z) \neq 0 \;\; \forall z \in \partial D$ and $\varphi(\xi) \neq 0 \;\; \forall\xi \in \partial M$ .

Our main result is:

Theorem. Conditions A. and B. below are equivalent

A. There exists a neighborhood $U$ of $\overline{M}$ in $\mathbb{C}^n$ and a holomorphic retraction $R : U \cap D \to M$ which is $C^\infty$ up to the boundary.

B. There exist functions $h_{ij} \in A^\infty(M)$ $i = 1,\cdots,k$ ; $j = k+1,\cdots,n$ so that

$$\varphi_j(\xi) = \sum_{i=1}^{k} \varphi_i(\xi) \cdot h_{ij}(\xi) \quad \forall\xi \in \partial M \; . \tag{2.1}$$

Remark.

a) We note that with our method we are not able to show that condition B. implies the existence of a global retraction $R : D \to M$ .

b) In the special case when $n = 2$ and $k = 1$ condition B. can be stated in a much nicer way; this is

C. The function $h(\xi) = \dfrac{\varphi_2(\xi)}{\varphi_1(\xi)}$ ; defined for all $\xi \in \partial M$ ; admits a holomorphic extension to $M$ .

Before we start with the proof of the theorem we give some examples.

Example 1.

If $D$ is any ball in $\mathbb{C}^n$ we can always construct such a retraction. In fact, by means of a linear transformation of $\mathbb{C}^n$ and the automorphisms of the unit ball we can always reduce the problem to the case $D = B$ = unit ball in $\mathbb{C}^n$ ; $M = B \cap \mathbb{C}^k$ and in this case the natural projection $\mathbb{C}^n \to \mathbb{C}^k$ provides the desired retraction.

The next example shows that the retraction $R$, in case it exists, is not necessarily unique.

Example 2.

Put $D = \{z \in \mathbb{C}^2 \,|\, |z_1|^2 + |1 - z_2|^2 < 2\}$ and $M = \{(z_1, z_2) \in \mathbb{C}^2 \,|\, z_2 = 0, \ |z_1| < 1\}$.

All the tangent planes to $D$ at points $\xi \in \partial M$ intersect at the point $(0, -1)$, so we can obtain a retraction $R$, by projecting into $M$ from the point $(0, -1)$. A simple computation gives $R_1(z_1, z_2) = \dfrac{z_1}{1 + z_2}$.

On the other hand define $R_2(z_1, z_2) = z_1 \cdot (1 - z_2)$. Then if $(z_1, z_2) \in D$ we have $|z_1|^2 + |1 - z_2|^2 < 2$ and so $2 \cdot |z_1| \cdot |1 - z_2| \le |z_1|^2 + |1 - z_2|^2 < 2$. Or $|z_1| \cdot |1 - z_2| < 1$. This means $R_2(z_1, z_2) \in M$ and $R_2$ is clearly a retraction, different from $R_1$. Moreover any convex combination of $R_1$ and $R_2$ will give again a retraction.

So far we have given examples where retractions do exist; with the help of condition C. it is not difficult to construct examples where a retraction fails to exist. One possibility is:

Example 3.

Take $D = B$ = unit ball in $\mathbb{C}^2$ ; then $M = U$ = unit disc in $\mathbb{C}$. A computation shows that in this case $h(\xi) = \dfrac{\varphi_2(\xi)}{\varphi_1(\xi)} \equiv 0 \quad \forall \xi \in \partial M$.

Now make a small perturbation of $D$ near the point $(-1, 0)$ in such a way that the convexity is preserved but now $h(\xi) = \dfrac{\varphi_2(\xi)}{\varphi_1(\xi)}$ is identically $0$ far away from $(-1,0)$ but different from $0$ near $(-1,0)$. Then $h$ cannot be the restriction to $\partial M$ of a function in $A^{\infty}(M)$ and consequently a retraction $R : D \to M$ fails to exist.

Example 4.

In $\mathbb{C}^2$ define $\rho(z_1, z_2) = z_1 \cdot \bar{z}_1 + z_2 \cdot \bar{z}_2 + \varepsilon \cdot (\bar{z}_1^2 \cdot z_2 + z_1^2 \cdot \bar{z}_2) - 1$. If $\varepsilon$ is chosen small enough $\rho$ is strictly convex in a neighborhood $D'$ of the unit ball in $\mathbb{C}^2$. Put $D = \{z \in D' \mid \rho(z) < 0\}$, then we have $M = D \cap \mathbb{C} = \{z_1 \in \mathbb{C} \,\big|\, |z_1| < 1\}$. Now for $\xi \in \partial M$ we have $h(\xi) = \frac{\varphi_2(\xi)}{\varphi_1(\xi)} = \frac{\bar{\xi}^2}{\xi} = \bar{\xi} = \frac{1}{\xi}$. Consequently condition C. is violated and a retraction $R : D \to M$ fails to exist.

Remark.

In example 4 the function $h(\xi) = \frac{1}{\xi}$ is holomorphic in a neighborhood of $\partial M$ in $\mathbb{C}$ and exactly as in the proof of B. $\Longrightarrow$ A., that will be given later, we can construct a neighborhood $U_1$, of $\partial M$ in $\mathbb{C}^2$ and a retraction $R_1 : D \cap U_1 \to \overline{M}$. Moreover, by using the natural projection $\mathbb{C}^2 \to \mathbb{C}$, we can find an open set $U_2$ and a retraction $R_2 : U_2 \cap D \to \overline{M}$ so that $\overline{M} \subset U_1 \cap U_2$. However these two retractions cannot be patched together.

Example 5.

In the case $k \geq 2$, if there exists a neighborhood $V$ of $\partial M$ in $\mathbb{C}^n$ and a retraction $R_1 : V \cap D \to \overline{M}$, the same argument as in the proof of A. $\Longrightarrow$ B. will show that there exist $h_{ij} \in C^{\infty}(\partial M)$ satisfying (2.1) in condition B. but with the $h_{ij}$'s holomorphic in $V \cap M$. Now by Hartog's Lemma $h_{ij} \in A^{\infty}(M)$ and condition B. is satisfied and according to our theorems there exists a neighborhood $U$ of $\overline{M}$ in $\mathbb{C}^n$ and a retraction $R : U \cap D \to \overline{M}$. This differs strongly from what was said in the remark after Example 4.

The next example gives a general method for constructing examples where retractions do exist.

Example 6.

Let $M = \{z' \in \mathbb{C}^k \mid \rho'(z') < 0\}$ be a strictly convex domain in $\mathbb{C}^k$ with $C^\infty$ boundary. Assume that $h_{ij}(z) \in A^\infty(M)$, $i = 1, \cdots, k$; $j = k+1, \cdots, n$ are given.

For $z = (z', z'') \in \mathbb{C}^k \times \mathbb{C}^{n-k} = \mathbb{C}^n$ define

$$\rho(z) = \rho'(z') + |z_{k+1}|^2 + \cdots + |z_n|^2 +$$

$$+ \varepsilon \cdot \sum_{\substack{i=1 \\ j=k+1}}^{\substack{n \\ k}} \left(\frac{\partial\rho'}{\partial z_i}(z')\cdot h_{ij}(z')\cdot z_j + \overline{\frac{\partial\rho'}{\partial z_i}(z')\cdot h_{ij}(z')\cdot z_j}\right)$$

Now $\rho$ is real valued and if $\varepsilon$ is chosen small enough $\rho$ is strictly convex. Put $D = \{z \in \mathbb{C}^n \mid \rho(z) < 0\}$; then $D \cap \mathbb{C}^k = M$ and for $(\xi, 0) \in M$ we obtain $\varphi_i(\xi) = \frac{\partial\rho}{\partial z_i}(\xi, 0) = \frac{\partial\rho'}{\partial z_i}(\xi)$ for $i = 1, \cdots, k$ and for $j = k+1, \cdots, n$

$$\varphi_j(\xi) = \frac{\partial\rho}{\partial z_j}(\xi, 0) = \varepsilon \cdot \sum_{i=1}^{k} \frac{\partial\rho'}{\partial z_i}(\xi)\cdot h_{ij}(\xi)$$

so condition B. is satisfied and by our theorem there exists a neighborhood $U$ of $\overline{M}$ in $\mathbb{C}^n$ and a retraction $R : U \cap D \to M$.

Example 7.

Let $D = \{z \in D' \mid \rho < 0\}$ be strictly convex in $\mathbb{C}^n$ and as before put $M = D \cap \mathbb{C}^k$. For $j = k+1, \cdots, n$ define

$$D_j = D \cap (\mathbb{C}^k \times \{0\} \times \cdots \times \{0\} \times \underset{\substack{\downarrow \\ j^{th}\ \text{place}}}{\mathbb{C}} \times \{0\} \times \cdots \times \{0\}).$$

Assume that there exist retractions $R_j : D_j \to M$. Then by the Theorem applied to each $D_j$ we can find functions $h_{ij} \in A^\infty(M)$

$i = 1, \cdots, k$ so that $\varphi_j = \sum_{i=1}^{k} \varphi_i \cdot h_{ij}$ on $\partial M$. Therefore condition B. is satisfied and we can find a neighborhood $U$ of $\overline{M}$ in $\mathbb{C}^n$ and a retraction $R : U \cap D \to M$.

## §3 Proof of A $\Longrightarrow$ B

In order to make the notation simpler we consider only the case $n = 2$ and $k = 1$, the proof in the general case being the same.

We assume that A. is satisfied and we must find a function $h = h_{12} \in A^{\infty}(M)$ so that $\varphi_2(\xi) = \varphi_1(\xi) \cdot h(\xi) \quad \forall \xi \in \partial M$.

In order to do this we consider the map $H : \mathbb{C}^2 \to \overline{M} \times \mathbb{C}$ given by $(z_1, z_2) \to (R(z_1, z_2), z_2)$.

Since $H/M = \text{id}$ and the complex Jacobian of $H$ at any point $(\xi, 0) \in \overline{M}$ is given by

$$JH(\xi, 0) = \det \begin{bmatrix} \frac{\partial R}{\partial z_1}(\xi, 0) & 0 \\ \frac{\partial R}{\partial z_2}(\xi, 0) & 1 \end{bmatrix} = 1$$

and $H$ defines a homeomorphism from a neighborhood of $\overline{M}$ in $\mathbb{C}^2$ to a neighborhood of $\overline{M}$ in $\mathbb{C}^2$.

Near $\overline{M}$, the function $\rho' = \rho \circ H^{-1}$ is a defining function for the image domain $H(D)$ and since $H(D) = \overline{M} \times \mathbb{C}$ the complex tangent plane to $H(D)$ at any point $(\xi, 0) \in \partial M \subset \overline{M} \times \mathbb{C}$ must be "vertical" ; this implies

$$\frac{\partial \rho'}{\partial z_2}(\xi, 0) = 0 \qquad \forall \xi \in \partial M . \tag{3.1}$$

On the other hand a simple computation using the chain rule shows

$$\frac{\partial \rho'}{\partial z_2}(\xi, 0) = \frac{\partial \rho}{\partial z_2}(\xi, 0) - \frac{\partial \rho}{\partial z_1}(\xi, 0) \cdot \frac{\partial R}{\partial z_2}(\xi, 0) . \tag{3.2}$$

But $\frac{\partial R}{\partial z_2}(\xi, 0) \in A^\infty(M)$, so setting $h(\xi) = \frac{\partial R}{\partial z_2}(\xi, 0)$ we obtain from (3.1) and (3.2)

$$\varphi_2(\xi) = \varphi_1(\xi)\cdot h(\xi) \qquad \forall \xi \in \partial M$$

as we wanted to show.

§4 Proof that B $\Longrightarrow$ A

We need now some extra notation.

Define

$$\emptyset(\xi, z) = \sum_{i=1}^{n} \varphi_i(\xi)\cdot(z_i - \xi_i) \quad \text{for} \quad \xi \in \partial M\,; \quad z \in \mathbb{C}^n$$

and

$$C = \{z \in \mathbb{C}^n \mid \operatorname{Re} \emptyset(\xi, z) < 0 \quad \forall \xi \in \partial M\}$$

We recall that for a fixed $\xi \in \partial M$, $\operatorname{Re}\emptyset(\xi, z) = 0$ is the equation of the real tangent plane to D at $(\xi, 0)$. Hence for a fixed $\xi \in \partial M$ $\{z \in \mathbb{C}^n \mid \operatorname{Re}\emptyset(\xi, z) < 0\}$ is a half plane containing D. and C is the intersection of these half planes as $\xi$ runs over $\partial M$. In particular C contains D.

We assume now that condition B. is satisfied, this is we assume the existence of $h_{ij} \in A^\infty(M)$ satisfying (2.1) and we will prove (4.1) below which is slightly stronger than A.

There exists a neighborhood U of $\overline{M}$ in $\mathbb{C}^n$ and a $C^\infty$ retraction $R : U \to \mathbb{C}^k$ so that $R|U \cap C : U \cap C \to \overline{M}$ is a holomorphic retraction. (4.1)

a) Construction of R :

Consider the map

$$F : \mathbb{C}^k \times \mathbb{C}^n \to \mathbb{C}^k$$

defined by $(\xi, z) \to (T_1(\xi, z), \cdots, T_k(\xi, z))$ where

$$T_i(\xi, z) = z_i - \xi_i + \sum_{j=k+1}^{n} h_{ij}(\xi)\cdot z_j \tag{4.2}$$

and the $h_{ij}$'s are the functions assumed to exist in condition B .

Then F is defined and $C^\infty$ in a neighborhood of $\overline{M} \times (\overline{M} \times \{0\})$ ; $F(\xi, (\xi, 0)) = 0$ , $\forall \xi \in \overline{M}$ and $d_\xi F(\xi, (\xi, 0)) = \text{id}$ $\forall \xi \in \overline{M}$ . So as a consequence of the implicit mapping theorem there exists a neighborhood U of $\overline{M}$ on $\mathbb{C}^n$ and a unique $C^\infty$ map $R = (R_1, \cdots, R_k) : U \to \mathbb{C}^k$ so that $R(\xi, 0) = \xi$ $\forall \xi \in \overline{M}$ and $F(R(z), z) = 0$ $\forall \xi \in U$ .

b) Properties of R :

If $R(z) \in M$ ; then R is holomorphic in a neighborhood of z and $R(\xi, 0) = \xi$ $\forall \xi \in \overline{M}$ . (4.3)

$T_i(\xi, z) = 0$ $\forall i = 1, \cdots, k \Longrightarrow R(z) = \xi$ (4.4)

There exist $C^\infty$ functions $f_{i\ell}(\xi, z)$ ; $g_{i\ell}(\xi, z)$ , $i, \ell = 1, \cdots, k$ so that (4.5)

$$R_i(z) = \xi_i + T_i(\xi, z) + \sum_{\ell=1}^{k} (f_{i\ell}(\xi, z)\cdot T_\ell(\xi, z) + g_{i\ell}(\xi, z)\cdot \overline{T_\ell(\xi, z)})$$

The functions $f_{i\ell}$ and $g_{i\ell}$ of (4.5) satisfy (4.6)

$$f_{i\ell}(\xi, (\xi, 0)) = g_{i\ell}(\xi, (\xi, 0)) = 0 \quad \forall \xi \in \overline{M} .$$

Properties (4.3) and (4.4) are immediate. As for the proof of (4.5) consider the maps

$$H_i : \mathbb{C}^k \times \mathbb{C}^n \to \mathbb{C} \qquad i = 1, \cdots, k$$

$$(\xi, z) \to R_i(z) - \xi_i - T_i(\xi, z) \quad .$$

Define

$$T : \mathbb{C}^k \times \mathbb{C}^n \to \mathbb{C}^k \times \mathbb{C}^n$$

$$(\xi, z) \to (\xi, (T_1(\xi, z), \cdots, T_k(\xi, z), z_{\ell+1}, \cdots, z_n))$$

$$(\eta, (\omega_1, \cdots, \omega_n)) = (\eta, \omega)$$

It is not difficult to check that $T$ defines a $C^\infty$ change of variables in a neighborhood of $\overline{M} \times (\overline{M} \times \{0\})$ on $\mathbb{C}^k \times \mathbb{C}^n$ .

Now $H_i \circ T^{-1}/\{(\eta, \omega) \in \mathbb{C}^k \times \mathbb{C}^n / \omega_1 = \cdots = \omega_k = 0\} \equiv 0$ and by a well known argument we can find functions $f'_{i\ell}(\eta, \omega)$ , $g'_{i\ell}(\eta, \omega)$ so that

$$H_i \circ T^{-1}(\eta, \omega) = \sum_{\ell=1}^{k} (f'_{i\ell}(\eta, \omega) \cdot \omega_\ell + g'_{i\ell}(\eta, \omega) \cdot \overline{\omega}_\ell) \quad .$$

Or

$$H_i(\xi, z) = \sum_{\ell=1}^{k} (f_{i\ell}(\xi, z) \cdot T_\ell(\xi, z) + g_{i\ell}(\xi, z) \cdot \overline{T_\ell(\xi, z)})$$

with $f_{i\ell} = f'_{i\ell} \circ T$

$$g_{i\ell} = g'_{i\ell} \circ T$$

Finally to prove (4.6) restrict the equation in (4.5) to $\overline{M} \times (\overline{M} \times \{0\})$ and differentiate with respect to $z_\ell$ (respectively $\overline{z}_\ell$) .

c) End of the proof assuming a Main Lemma .

We note first that if we can find a neighborhood $V$ of $\partial M$ in $\mathbb{C}^n$ so that $R/C \cap V : C \cap V \to M$ is a retraction we are done. This is because the continuity of $R$ and the fact that $R/\overline{M} = \text{id}$ allow us to construct the desired neighborhood $U$ once we have $V$ .

Now fix $\xi \in \partial\overline{M}$ . The Taylor series expansion of $\rho/\mathbb{C}^k$ about the point $\xi$ is

$$\rho(\omega_1, \cdots, \omega_k) = 2 \operatorname{Re} \sum_{i=1}^{k} \varphi_i(\xi) \cdot (\omega_i - \xi_i) + \tag{4.7}$$

$$H(\xi)(\omega - \xi \, , \quad \omega - \xi) + 0(|\omega - \xi|^2)$$

where $H(\xi)(\ ,\ )$ denotes the quadratic form associated with the real Hessian of $\rho$ .

The strict convexity of $\rho$ and the compactness of $\partial M$ allows us to say that $\omega$ is close enough to $\xi$ ; say $|\omega - \xi| < \varepsilon_1$ ; then

$$\rho(\omega) \leq 2 \operatorname{Re} \sum_{i=1}^{k} \varphi_i(\xi)(\omega_i - \xi_i) + K \cdot |\xi - \omega|^2 \tag{4.8}$$

where $K$ and $\varepsilon_1$ are independent of $\xi \in \partial M$ .

Now if we observe that (2.1) in condition B. implies

$$\emptyset(\xi, z) = \sum_{i=1}^{k} \varphi_j(\xi) \cdot T_i(\xi, z) \quad \forall \xi \in \partial M , \tag{4.9}$$

using (4.5) , (4.6) , (4.8) and (4.9) we obtain

$$\rho(R(z)) \leq 2 \operatorname{Re} \emptyset(\xi, z) + A(\xi, z) \cdot |T(\xi, z)| + K \cdot |T(\xi, z)|^2 \tag{4.10}$$

whenever $|\xi - z| < \varepsilon_2$ , for a suitable $\varepsilon_2$ , and where $|T(\xi, z)| = |T_1(\xi, z)| + \cdots + |T_k(\xi, z)|$ and $A(\xi, z)$ is a $C^\infty$ function that satisfies $A(\xi, (\xi, 0)) = 0 \quad \forall \xi \in \overline{M}$ .

We need now:

<u>Main Lemma</u>.

Given $\varepsilon > 0$ there exists a neighborhood $U_\varepsilon$ of $\partial M$ in $\mathbb{C}^n$ so that: $\forall z \in U_\varepsilon$ there exists $\xi(z) = \xi \in \partial M$ satisfying

$$|\xi(z) - z| < \varepsilon \tag{4.11}$$

$$\emptyset(\xi(z), z) \in \mathbb{R} \tag{4.12}$$

$$T_i(\xi(z), z) = \frac{\overline{\varphi_i(\xi)}}{|\varphi(\xi)|^2} \cdot \emptyset(\xi(z), z) \qquad 1 \leq i \leq k \tag{4.13}$$

We will assume for the moment the lemma above. In order to finish our proof pick now $\varepsilon > 0$ , $\varepsilon < \varepsilon_2$ and so that $|\xi - z| < \varepsilon \Longrightarrow \big| A(\xi, z) + K \cdot |T(\xi, z)| \big| < 1$ . By the above lemma there exists a neighborhood $U_\varepsilon$ of $\partial M$ with the above stated properties.

Now if $z \in U_\varepsilon \cap C$ , by using (4.10) , (4.12) and (4.13) , we get

$$\rho(R(z)) \leq -2\cdot|\emptyset(\xi(z),z)| + |\emptyset(\xi(z),z)| = -|\emptyset(\xi(z),z)| < 0 \quad .$$

This means $R(z) \in M$ whenever $z \in U_\varepsilon \cap C$ .

This finishes the proof.

d) Proof of the Main Lemma .

Fix $\xi^* \in \partial M$ . We are going to construct a $C^\infty$ change of variables in $\mathbb{C}^n$ of the form $H : \mathbb{R}^{2n} \to \mathbb{C}^n$ so that:

The domain of $H$ is a cube centered at $0$ , $H(0) = \xi^*$ and the range has diameter less then $\varepsilon$ (4.14)

$$H(\mathbb{R}^{2k-1} \times \{0\} \times \cdots \times \{0\}) \subset \partial M \tag{4.15}$$

$$T_i(H(t_1,\cdots,t_{2k-1},0,\cdots,0)\,;\;H(t)) = \tag{4.16}$$

$$= \frac{\overline{\varphi_i(H(t))}}{|\varphi(H(t))|^2}\cdot\eta(t) \qquad 1 \leq i \leq k$$

where $\eta(t) \in \mathbb{R}$ .

We observe that once we have constructed a change of variables $H$ as above it is easy to obtain the desired neighborhood $U_\varepsilon$ of the lemma. In order to see this take a finite covering of $\partial M$ by ranges of the above type of maps. Call the union of this ranges $U_\varepsilon$ . Now any point $z \in U_\varepsilon$ will be of the form $z = H(t)$ for some $H$ and some $t$ . If we define $\xi(z) = H(t_1,\cdots,t_{2k-1},0,\cdots,0)$ we can see that (4.11) is implied by (4.14) ; $\xi(z) \in \partial M$ because of (4.15) and since by (4.16)

$$\emptyset(\xi(z),t) = \sum_{i=1}^{k} \varphi_i(\xi(z))\cdot T_i(\xi(z),z) =$$

$$= \sum_{i=1}^{k} \frac{\varphi_i(\xi(z))\cdot\overline{\varphi_i(\xi(z))}}{|\varphi(t)|^2}\cdot\eta(t) = \eta(t) \in \mathbb{R}$$

we have (4.12) and (4.13) .

Now the only thing that remains to be done is to construct $H$. In order to do this let

$$\xi(t') : \mathbb{R}^{2k-1} \to \partial M$$

which maps: $t' \to (u_1(t') + iv_1(t'), \cdots, u_k(t') + iv_k(t')) = \xi(t')$ be a parametrization of $\partial M$ with $\xi(0) = \xi^*$.

We will put $t = (t', t_{2k}, t'') \in \mathbb{R}^{2k-1} \times \mathbb{R} \times \mathbb{R}^{2(n-k)} = \mathbb{R}^{2n}$ and $z_j = t_{2j-1} + it_{2j}$.

Consider the function

$$F : \mathbb{R} \times \mathbb{R}^{2n} \to \mathbb{R}$$

defined by

$$F(\eta, t) = \rho(\alpha_1(\eta, t), \beta_1(\eta, t), \cdots, \alpha_k(\eta, t), \beta_k(\eta, t), t'') + t_{2k+1}$$

where

$$\alpha_i(\eta, t) = u_i(t') + \frac{\frac{\partial \rho}{\partial x_i}(\xi(t'))}{|\varphi(\xi(t'))|^2} \cdot \eta - \mathrm{Re}\left[\sum_{j=k+1}^{n} h_{ij}(\xi(t')) \cdot z_j\right]$$

$$\beta_i(\eta, t) = v_i(t') + \frac{\frac{\partial \rho}{\partial y_i}(\xi(t'))}{|\varphi(\xi(t'))|^2} \cdot \eta - \mathrm{Im}\left[\sum_{j=k+1}^{n} h_{ij}(\xi(t')) \cdot z_j\right]$$

We can check that the hypothesis of the implicit function theorems are satisfied and so there exists a neighborhood of $0$ in $\mathbb{R}^{2n}$ and a $C^\infty$ function $\eta : \mathbb{R}^{2n} \to \mathbb{R}$ so that $\eta(t', 0, \cdots, 0) = 0 \quad \forall t' \in \mathbb{R}^{2k-1}$ and $F(\eta(t), t) = 0 \quad \forall t$.

Now we define

$$H : \mathbb{R}^{2n} \to \mathbb{C}^n$$

$$t \to (p_1(t) + iq_1(t), \cdots, p_k(t) + iq_k(t), z_{\ell+1}, \cdots, z_n)$$

where

$$p_i(t) = \alpha_i(\eta(t);t)$$

$$q_i(t) = \beta_i(\eta(t);t) \quad .$$

Using the chain rule, or otherwise, one obtains

$$\frac{\partial p_i}{\partial t_j}(0) = \frac{\partial u_i}{\partial t_j}(0), \quad \frac{\partial q_i}{\partial t_j}(0) = \frac{\partial v_i}{\partial t_j}(0) \quad \text{for } j = 1, \cdots, 2k-1;$$

$$\frac{\partial p_i}{\partial t_{2k}}(0) = -\frac{\frac{\partial \rho}{\partial x_i}(0)}{|\varphi(0)|^2}, \quad \frac{\partial q_i}{\partial t_{2k}}(0) = -\frac{\frac{\partial \rho}{\partial y_i}(0)}{|\varphi(0)|^2}$$ and so the Jacobian of $H$

at $0$ is of the form

$$-\frac{1}{|\varphi(0)|^2} \det \begin{bmatrix} \vec{t}_1 & \cdots & \vec{t}_{2k-1} & \vec{\varphi} \\ \downarrow & & \downarrow & \downarrow \\ & & & \end{bmatrix}$$

where $\vec{t}_1, \cdots, \vec{t}_{2k-1}$ are vectors that generate the real tangent space to $\partial M$ at $0$ and $\varphi$ is the gradient of $\rho/\mathbb{C}^k$ at $\xi^*$ and hence perpendicular to that tangent plane. Consequently the determinant is nonzero and $H$ is a change of variables. Property (4.14) can be obtained by further reducing the domain of $H$ if necessary, (4.15) is immediate from the construction and (4.16) is just algebra.

## Bibliography


[1] F. Docquier and H. Grauert; Levisches Problem und Rungesher Satz fur Teil gebiete Steinscher Mannifaltigkeiten, Math. Ann. 140 (1960), 94-123.

[2] H. Rossi; A Docquier-Grauert lemma for strongly pseudoconvex domains in complex manifolds, Rocky Mountain Journal of Math., Vol. 6, No. 1 (1976), 171-176.

[3] W. Rudin; Function theory in polydiscs. Benjamin (1969).

# 4
# A Boundary Canonical Transformation for a Class of Operators with Double Characteristics

GUNTHER A. UHLMANN / Department of Mathematics, Massachusetts Institute of Technology, Cambridge, Massachusetts

## Introduction

Every pseudodifferential operator $P$ on a $C^\infty$ manifold $X$ of dim $n$ with simple real characteristics can be conjugated by a Fourier integral operator to a pseudodifferential operator with principal symbol $D_{x_1}$. (See [1]). The conjugation involves two main steps. First, an homogeneous symplectic transformation $\chi$: $T^*(X) \longrightarrow T^*(R^n)$ is constructed so that $p \circ \chi = \xi_1$ where $p$ is the principal symbol of $P$ and $\xi_1$ is the principal symbol of $D_{x_1}$. Second, Egorov's theorem says that the principal symbol of $BPA$ where $A$ (resp. $B$) are Fourier integral operators associated to $\chi$ (resp. $\chi^{-1}$) is $p \circ \chi = \xi_1$. This reduction to a "normal form" simplifies the construction of microlocal parametrices for operators of real principal type and allows a simple proof of propagation of singularities for such operators (see [1] for more details). "Normal forms" are also known for a class of hyperbolic equations with double characteristics (see [4], for instance).

However the method outlined above has some drawbacks. For instance it is in general not possible to apply it to solve boundary value problems or initial value problems since the symplectic transformation constructed does not preserve the boundary or the initial surface. Recently Melrose has developed a theory of operators commuting with restriction to the boundary similar to the theory of Fourier integral operators. (See [2].) This theory can be applied to prove propagation of singularities for boundary value problems for equations of principal type. In this note we solve the Cauchy problem for hyperbolic equations with double involutive characteristics by conjugating with an operator associated to a symplectic transformation preserving the initial surface. We are reduced then to an operator with constant coefficient principal part. This class of equations is interesting to study because of the "wedge refraction" phenomenon which shows up at the double characteristic set (see [4]).

A note about the presentation of the paper. In section 1 we make our assumptions. In section 2 we reduce the operator to a system. In section 3 we reduce the system to a simpler one. In section 4 we reduce the system further to a system whose principal part has constant coefficients. Finally in section 5 we finalize the construction of a microlocal parametrix using the results of [4].

I would like to thank R. Melrose and L. Nirenberg for helpful conversations.

## 1. Assumptions

Let $X \subseteq R^{n-1}$ be an open set, $n \geq 2$. Let $Y = R \times X$. Variables in $T^*(R^n)$ will be denoted by $((t,\tau), (x,\xi)) \in T^*(R) \times T^*(X)$. We will consider operators of the form

$$P = (D_t - \lambda_1(t,x,D_x))(D_t - \lambda_2(t,x,D_x)) + Q \tag{1.1}$$

where $\lambda_i(t,.,.) \in L^1(X)$ varying smoothly with $t$, $i = 1, 2$, $Q \in L^1(Y)$.

All pseudodifferential operators are assumed to be properly-supported and classical. If $P$ is a pseudodifferential operator we will denote by $p$ its principal symbol. $\lambda_i(t,x,\xi)$ are real-valued smooth homogeneous functions of degree 1; $d\lambda_1$, $d\lambda_2$ and the cone axis are linearly independent at $\lambda_1 = \lambda_2$. We assume

$$\{\tau - \lambda_1(t,x,\xi), \tau - \lambda_2(t,x,\xi)\} = 0 \quad \text{on} \quad \tau = \lambda_1 = \lambda_2 \tag{1.2}$$

and

$$q = 0 \quad \text{on} \quad \tau = \lambda_1 = \lambda_2. \tag{1.3}$$

Condition (1.2) is equivalent to the Levi condition (cf. [4]), which is necessary for the Cauchy problem to be well-posed for operators of the form (1.1).

## 2. Reduction to an hyperbolic system

Let us consider $P$ of the form (1.1) and satisfying (1.2) and (1.3). First we reduce $P$ to the case $Q$ of order 0.

Proposition 2.1. We can choose the terms of order 0 in the full symbol of $\lambda_1(t,x,D_x), \lambda_2(t,x,D_x)$ so that

$$P = (D_t - \lambda_1(t,x,D_x))(D_t - \lambda_2(t,x,D_x)) + T$$

with $T \in L^0(Y)$.

Proof: Let $p_1 \in S^1(T^*(Y) - \{0\})$ be the term of order one in the asymptotic expansion of the full symbol of $p$. Comparing terms of order 1 in

$$P = (D_t - \lambda_1(t,x,D_x))(D_t - \lambda_2(t,x,D_x)) + Q \tag{2.2}$$

we obtain

$$p_1 = -\lambda_1^0(\tau - \lambda_2) - D_t\lambda_2 + \sum_{j=1}^{n-1} \frac{\partial\lambda_1}{\partial\xi_j} D_{x_j}\lambda_2 - (\tau - \lambda_1)\lambda_2^0 + q$$

since $q$ vanishes when $\tau = \lambda_1 = \lambda_2$, we choose $\lambda_1^0$, $\lambda_2^0$, so that

$$q - \lambda_1^0(\tau - \lambda_2) - \lambda_2^0(\tau - \lambda_1) = 0, \tag{2.3}$$

proving the proposition. Q.E.D.

Let us consider the system

$$K = \begin{pmatrix} D_t - \lambda_1(t,x,D_x) & 0 \\ 0 & D_t - \lambda_2(t,x,D_x) \end{pmatrix} + \begin{pmatrix} 0 & T \\ -Id & 0 \end{pmatrix} \tag{2.4}$$

Let $N(t,x,D_t,D_x) = \begin{pmatrix} 0 & T \\ -Id & 0 \end{pmatrix}$. Note that (2.4) is the "natural" system associated to (2.2).

Using arguments similar to Proposition 3.2, Section 3 in [4] developing in Taylor series around $\tau = \lambda_1(t,x,\xi)$ in the first column and $\tau = \lambda_2(t,x,\xi)$ in the second column we can construct $C \in L^0(Y)$ elliptic near $(0,x_0,0,\xi_0)$ and $N(t,x,D_x) \in L^0(X)$ varying smoothly with $t$ such that

$$CK = C\left[\begin{pmatrix} D_t - \lambda_1(t,x,D_x) & 0 \\ 0 & D_t - \lambda_2(t,x,D_x) \end{pmatrix} + N(t,x,D_x)\right]. \tag{2.5}$$

## 3. Reduction to a simpler system

Here we reduce the system

$$L = \begin{pmatrix} D_t - \lambda_1(t,x,D_x) & 0 \\ 0 & D_t - \lambda_2(t,x,D_x) \end{pmatrix} + N(t,x,D_x) \tag{3.1}$$

as in (2.5) to the case $\lambda_1 = 0$.

Let us take $((0,\tau_o), (x_o,\xi_o)) \in T^*(R) \times (T^*(X) - \{0\})$, such that $\tau_o = \lambda_1(0,x_o,\xi_o) = \lambda_2(0,x_o,\xi_o) = 0$. Let $\phi \in C^\infty(R \times (T^*X - \{0\}))$ be homogeneous of degree 1 in $\xi$ satisfying

$$\left\{\begin{aligned} \frac{\partial\phi}{\partial t} &= \lambda_1(t,x,d_x\phi) \\ \phi(0,x,\xi) &= \langle x,\xi\rangle \end{aligned}\right\} . \tag{3.2}$$

Let us take

$$\Phi(t,x,\tau,\xi) = \phi(t,x,\xi) + t\tau . \tag{3.3}$$

Note that

$$\left\{\begin{aligned} \Phi(0,x,\tau,\xi) &= \langle x,\xi\rangle \\ d_t\Phi(t,x,\tau,\xi) &= \lambda_1(t,x,d_x\phi) + \tau \\ d_x\Phi(t,x,\tau,\xi) &= d_x\phi(t,x,\xi) \\ d_\tau\Phi(t,x,\tau,\xi) &= t \\ d_\xi\Phi(t,x,\tau,\xi) &= d_\xi\phi(t,x,\xi). \end{aligned}\right. \tag{3.4}$$

Let us consider the canonical transformation $\chi$ from a conic neighborhood of $(0,0,x_o,\xi_o)$ to a conic neighborhood of $(0,0,x_o,\xi_o)$ defined by

$$\chi(t,x,d_t\Phi(t,x,\tau,\xi),\ d_x\Phi(t,x,\tau,\xi)) = (d_\tau\Phi,d_\xi\Phi,\tau,\xi). \tag{3.5}$$

Note that $\chi$ preserves the surface $t = 0$. Let us consider $A\colon\ C_o^\infty(R^n) \longrightarrow C^\infty(Y)$ defined by

$$Af(t,x) = \int e^{i(\phi(t,x,\xi) + t\tau)} a(t,x,\xi)\hat{f}(\tau,\xi)d\tau d\xi \tag{3.6}$$

with $\phi$ as in (3.3), $a \in S^o(Y \times R^{n-1})$, $a = 1$ in a conic neighborhood of $(0,x_o,\xi_o)$.

$A$ is a Fourier integral operator away from points of the form $((t,x,\tau,0);\ (\bar{t},\bar{x},\bar{\tau},0))$. Also $((0,x_o,0,\xi_o),\ (0,x_o,0,\xi_o))$ is a non-characteristic point for $A$.

Let $B \in I^0(R^n,(R \times X),\Gamma')$ be a microlocal inverse for $A$, where $\Gamma$ is a closed conic subset of the graph of $\chi$, (and $B$ of the same form as $A$), i.e.,

$$\begin{cases} (0,x_o,0,\xi_o) \notin WF(AB - Id) \\ (0,x_o,0,\xi_o) \notin WF(BA - Id). \end{cases} \tag{3.7}$$

Note that $A$ commutes by restriction to the $t = 0$ surface with a pseudodifferential operator in $R^{n-1}$, i.e.

$$\gamma_o A = \tilde{A}\gamma_o \quad \text{where} \quad \tilde{A} \in L^o(R^{n-1}). \tag{3.8}$$

$\tilde{A}$ elliptic near $(x_o,\xi_o)$ and $\gamma_o$ denotes the restriction to the $t = 0$ surface.

<u>Proposition 3.9.</u> The principal symbol of $\tilde{L}$ with $\tilde{L} = BLA$ is

$$\tilde{\ell} = \begin{pmatrix} \tau & 0 \\ 0 & \tau - \tilde{\lambda}(t,x,\xi) \end{pmatrix}$$

in a conic neighborhood of $(0,x_o,0,\xi_o)$, $\tilde{\lambda} \in C^\infty(R \times T^*(R^{n-1}) - \{0\})$ is a real-valued homogeneous function of degree $1$ in $\xi$.

<u>Proof:</u> We have to show

$$\ell(t,x,d_t\Phi,d_x\Phi) = \tilde{\ell}(t,d_\xi\Phi,\tau,\xi). \tag{3.10}$$

Because of (3.4) the left hand side of (3.10) is equal to

$$\begin{pmatrix} \tau & 0 \\ 0 & \tau + \lambda_1(t,x,d_x\phi) - \lambda_2(t,x,d_x\phi) \end{pmatrix} .$$

The right hand side of (3.10) is equal to

$$\begin{pmatrix} \tau & 0 \\ 0 & \tau - \tilde{\lambda}(t,x,\xi) \end{pmatrix} .$$

Thus we define

$$\tilde{\lambda}(t,d_\xi\phi,\xi) = \lambda_2(t,x,d_x\phi) - \lambda_1(t,x,d_x\phi) \tag{3.11}$$

Q.E.D.

Remark: The operator $L$ is not a pseudodifferential operator in $R \times X$ since we have

$$WF'L \subseteq \Delta \cup \{(t,x,\tau,0),(t,\bar{x},\tau,0)) \in T^*(R \times X) \times R \mid \tau \neq 0\}$$

where $\Delta$ is the diagonal in $(T^*(R\times X) \times T^*(R\times X)) - \{0\}$. However, since $A$ is a Fourier integral operator near $((0,x_o,0,\xi_o),(0,x_o,0,\xi_o))$, $L$ is a pseudodifferential operator near the same point $(\xi_o \neq 0)$ and $B$ is a Fourier integral operator near $((0,x_o,0,\xi_o),\ (0,x_o,\xi_o))$ we get that $BLA$ is a pseudodifferential operator. Since $B(t,x,D_x)$ and $A(t,x,D_x)$ don't depend on $D_t$ we get

$$BLA = \begin{pmatrix} D_t & 0 \\ 0 & D_t - \tilde{\lambda}(t,x,D_x) \end{pmatrix} + \tilde{N}(t,x,D_x) \tag{3.12}$$

in a conic neighborhood of $(o,x_o,0,\xi_o)$ with $\tilde{N}(t,.,.) \in L^0(R^{n-1})$ varying smoothly with $t$ and $\tilde{\lambda}(t,.,.)$ varying smoothly with $t$ with principal symbol $\tilde{\lambda}(t,x,\xi)$ as in (3.11). Let us suppose we can construct an operator $E\colon\ C_o^\infty(R^{n-1}) \longrightarrow C^\infty(R^n)$ s.t.

$$\begin{cases} \widetilde{LE} \in C^\infty(R^n \times R^{n-1}) \\ \gamma_o E = J \\ WF(AB - Id) \circ \chi^{-1} \circ WF'E = \phi \end{cases} \tag{3.13}$$

with $J \in L^0(R^{n-1})$, $\sigma(J) = 1$ in a conic neighborhood of $(x_o,\xi_o)$ and with essupp $\sigma(J)$ in another conic neighborhood of $(x_o,\xi_o)$.

Thus we have

$$BLAE \in C^\infty(R^n \times R^{n-1})$$

$$\gamma_o E = J \quad .$$

Let $\tilde{E} = AE$. Because of (3.8) and (3.13)

$$\gamma_o\tilde{E} = \gamma_o AE = \tilde{A}\gamma_o E = \tilde{A}J.$$

Thus using (3.13) again we obtain

$$\begin{cases} LE \in C^\infty((R \times X) \times R^{n-1}) \\ \\ \gamma_o E = D, \ D \in L^0(R^{n-1}) \end{cases} \tag{3.14}$$

with $\sigma(D) = 1$ in a conic neighborhood of $(x_o,\xi_o)$ and $\sigma(D)$ is the full symbol of D.

Since we can find such an $\tilde{E}$ for any $(x_o,\xi_o)$, $\xi_o \neq 0$ and we have pseudodifferential partitions of the unity we can find an operator $F: C_o^\infty(X) \longrightarrow C^\infty(Y)$ s.t.

$$\begin{cases} LF \in C^\infty((R \times X) \times R^{n-1}) \\ \gamma_o F = \text{Id mod. } L^{-\infty}(R^{n-1}) \end{cases} \tag{3.15}$$

Thus in this paragraph we have reduced the problem of constructing a parametrix for the Cauchy problem for L to construct a parametrix E for $\tilde{L}$ satisfying (3.13).

## 4. Reduction to an hyperbolic system with constant coefficients in the principal part

We have

$$\tilde{L} = \begin{pmatrix} D_t & 0 \\ 0 & D_t - \tilde{\lambda}(t,x,D_x) \end{pmatrix} + \tilde{N}(t,x,D_x) \tag{4.1}$$

The involutive condition (1.2) says

$$\frac{\partial\tilde{\lambda}}{\partial t} = 0 \ \text{ on } \ \tilde{\lambda} = 0. \tag{4.2}$$

Our intention in this paragraph is to reduce the construction of an operator E as in (3.13) to construct a parametrix for the Cauchy problem for an operator of the form

$$K = \begin{pmatrix} D_t & 0 \\ 0 & D_t - D_{x_1} \end{pmatrix} + H(t,x,D_x) \tag{4.3}$$

$H(t,.,.) \in L^0(R^{n-1})$ varying smoothly with $t$, and apply the construction of a parametrix for the Cauchy problem for systems of the form (4.3) given in great detail in [4].

The idea in making such a reduction consists in constructing a canonical transformation preserving the $t = 0$ surface and getting to the case where $\frac{\partial\tilde{\lambda}}{\partial t} = 0$ in a neighborhood of $(0,x_o,0,\xi_o)$.

Proposition 4.4. There exists a homogeneous canonical transformation $\chi$ from a conical neighborhood V of $(0,x_o,0,\xi_o)$ to another conical neighborhood of $(0,x_o,0,\xi_o)$ such that $\chi(t,x,\tau,\xi) = (t',x',\tau',\xi')$; $t' = 0$ on $t = 0$, $\sigma' = 0 \Leftrightarrow \sigma = 0$ with $\sigma = \tau - \tilde{\lambda}(t,x,\xi)$ and $\sigma' = \tau' - \xi_1'$; $\tau' = 0 \Leftrightarrow \tau = 0$.

Proof: We have that $\tilde{\lambda}(t,x,\xi) = e^{\int_o^t a(s,x,\xi)ds}\,\tilde{\lambda}(0,x,\xi)$ since with $a \in C^\infty(R \times T^*(R^{n-1}) - 0)$ homogeneous of degree 0 since

$$\frac{\partial\tilde{\lambda}}{\partial t} = a(t,x,\xi)\,\tilde{\lambda}(t,x,\xi).$$

Then $$\frac{\tilde{\lambda}(0,x,\xi)}{\tilde{\lambda}(t,x,\xi)} = e^{-\int_o^t a(s,x,\xi)} \in C^\infty(R \times T^*(R^{n-1}) - 0).$$

We choose $\sigma'$ by

$$\sigma' = \frac{\tilde{\lambda}(o,x,\xi)}{\tilde{\lambda}(t,x,\xi)}\,(\tau - \tilde{\lambda}(t,x,\xi)). \tag{4.5}$$

When $\tau = \tilde{\lambda}(\sigma = 0)\sigma' = 0$ and vice versa. Furthermore

$$H_\tau \sigma' = 0 \quad \text{on} \quad \tau = 0. \tag{4.6}$$

Let us choose $\tau'$ by solving

$$\begin{cases} H_{\sigma'}\tau' = 0 \\ \tau' = \tau \quad \text{on} \quad t = 0, \end{cases} \tag{4.7}$$

(4.7) has a solution since $H_{\sigma'}$ is transversal to the $t = 0$ surface.

We have, because of 4.7, that $\tau' = 0$ on $\tau = t = 0$. Since $H_{\sigma'}$ is transversal to $t = 0$ and because of (4.6) $H_{\sigma'}$ is tangent to $\tau = 0$ we get that $\tau' = 0$ on $\tau = 0$. Since $\tau' = \tau + t\tau k$, $\tau' = 0 \Rightarrow \tau = 0$ for $t$ near $0$.

Let us choose $t'$ so that

$$\begin{cases} H_{\sigma'}t' = 1 \\ t' = 0 \quad \text{on} \quad t = 0. \end{cases} \tag{4.8}$$

We certainly can solve (4.8) because $H_{\sigma'}$ is transversal to $t = 0$.

Claim:

$$\{\tau', t'\} = 1. \tag{4.9}$$

Proof: Using Jacobi's identity, (4.7) and (4.8) we obtain

$$\{\{\tau', t'\}, \sigma'\} = 0. \tag{4.10}$$

Since $\tau' = \tau + tk$, $t' =$ at with a, k $C^\infty$ functions because of the initial conditions in (4.7) and (4.8), we have

$$\{\tau', t'\} = a \quad \text{on} \quad t = 0.$$

Also

$$\{\tau', t\} = a \quad \text{on} \quad t = 0.$$

Thus we conclude that

$$\{\tau', t'\} = 1 \quad \text{on} \quad t = 0; \tag{4.11}$$

(4.11) together with (4.10) proves (4.9).

We can always assume without loss of generality that $\lambda(0,x,\xi) = \xi_1$, (taking a canonical transformation in the $(x,\xi)$ variables transforming $\lambda(0,x,\xi) \longrightarrow \xi_1$ and tensoring with the identity in the $(t,\tau)$ variables). Let us put $\sigma' = \tau' - \xi_1'$.

The construction of the remaining symplectic coordinates $x_j', j = 1,\ldots,n$, $\xi_j', j = 2,\ldots, n$ is done by solving

$$\begin{cases} H_{\sigma'} x_j' = \delta_{1j} & j = 1,\ldots, n \\ x_j' = x_j & \text{on} \quad t = 0 \\ H_{\sigma'} \xi_j' = 0 & j = 2,\ldots, n \\ \xi_j' = \xi_j & \text{on} \quad t = 0. \end{cases} \tag{4.12}$$

Using Jacobi's identity, Equations (4.12), (4.9), (4.8), (4.7) and the expression for $\sigma'$ (4.5) we get that

$$\{\tau', t\} = 1, \ \{\tau', \xi_1'\} = 0, \ \{\xi_1', t'\} = 0$$

$$\{x_j', \xi_k'\} = -\delta_{jk}, \ j,k = 1,\ldots,n$$

$$\{\xi_j', \xi_k'\} = 0, \ j,k = 1,\ldots,n,$$

$t' = 0$ on $t = 0$, $\tau' = 0$ on $\tau = 0$, $\sigma' = 0$ on $\sigma = 0$, proving the proposition. Q.E.D.

Remark:

$$\frac{\partial \xi_1}{\partial \tau} = -\{t, \xi_1'\} = \{\sigma' - \tau', t\} = 0 \quad \text{on} \quad t = 0.$$

Since $x_j' = x_j$, $j = 1,\ldots,n$, $\xi_j' = \xi_j$, $j = 2,\ldots,n-1$ on $t = 0$, then

$$\frac{\partial x_j'}{\partial \tau} = 0 \quad \text{on} \quad t = 0, j = 1,\ldots,n$$

$$\frac{\partial \xi_j}{\partial \tau} = 0 \quad \text{on} \quad t = 0, j = 2,\ldots,n-1.$$

Thus we conclude that $(x',\xi')$ are symplectic coordinates on $t = 0$ and the canonical transformation $\chi$ constructed in Proposition 2.1 is a boundary canonical transformation in the sense of [2].

Let $\phi(t,x,\tau,\xi)$ be a phase function generating $\chi$, i.e.

$$\chi(t,x,d_t\phi,d_x\phi) = (d_\tau\phi,d_\xi\phi,\tau,\xi).$$

Since $\chi$ preserves the $t = 0$ surface we have $d_\tau\phi = 0$ on $t = 0$. Then

$$\phi(t,x,\tau,\xi) = \phi(t,x,0,\xi) + t\tau f(t,x,\tau,\xi). \tag{4.13}$$

Let us consider the operator

$$A: \; C_0^\infty(R^n) \longrightarrow C^\infty(R^n)$$

$$Af(t,x) = \int e^{i\Phi(t,x,\tau,\xi)} a(t,x,t\tau,\xi)\hat{f}(\tau,\xi)d\tau d\xi. \tag{4.14}$$

We choose $a \simeq \sum_{j=0} a_{-j}$ with $a_{-j}$ homogeneous of degree $-j$ for $|(\xi,\tau)|$ large. We choose $a(t,x,\tau,\xi) = 1$ near $(0,x_o,0,\xi_o)$. A is an elliptic Fourier integral operator away from $\xi = 0$.

Let us take B an elliptic Fourier integral operator microlocal inverse of A at $(0,x_o,0,\xi_o)$ with $WFB \subset \Gamma$, $\Gamma$ closed subset of graph $\chi$. Let $K = B\tilde{L}A$. We have

$$K = \begin{pmatrix} D_t & 0 \\ 0 & D_t - D_{x_1} \end{pmatrix} + H(t,x,D_x) \tag{4.15}$$

near $(0,x_o,0,\xi_o)$, with $H(t,.,.) \in L^0(R^{n-1})$ varying smoothly with $t$.

<u>Remark</u>: Strictly speaking we would get

$$K = \begin{pmatrix} D_t & 0 \\ 0 & D_t - D_{x_1} \end{pmatrix} + H(t,x,D_x,D_t).$$

Applying the same construction as in [4], Section 3, Proposition 3.2, developing in Taylor series around $\tau = 0$ in the first column and $\tau = \xi_1$ in the second column, we can construct an operator $C \in L^0(R^n)$ elliptic near $(0,x_o,0,\xi_o)$ s.t.

$$CK = C \left[ \begin{pmatrix} D_t & 0 \\ 0 & D_t - D_{x_1} \end{pmatrix} + H(t,c,D_x) \right] \quad \text{mod. } L^{-\infty}(R^n).$$

We know from (4.13) and (4.14) that

$$\gamma_o A = \tilde{A}\gamma_o \quad \text{with} \quad \tilde{A} \in L^0(R^{n-1}) \quad \text{elliptic}$$

near $(x_o,\xi_o)$, because $\phi(0,x,0,\xi)$ is a phase function defining the canonical transformation generating $\chi$ on $t = 0$ which is the identity.

Using arguments similar to Section 2, in order to construct an operator $E$ as in (3.13) for $\tilde{L}$, we have to construct a similar one for $K$ with the new $A,B,\chi$ defined above, i.e.,

$$\begin{cases} KE \in C^\infty(R^n \times R^{n-1}) \\ \gamma_o E = J \\ WF(AB - Id) \circ \chi^{-1} \circ WF'E = \phi \end{cases} \tag{4.16}$$

## 5. Construction of a parametrix for the Cauchy problem for K

We are not going to give here too many details since the construction was given in great detail in [4], Section 4. Also a global construction was given in [3]. We recall that the parametrix $E$ for the Cauchy problem constructed in [4] has the form $E = E_1 + E_2 + E_3$ with

$$E_1 f(t,x) = \int e^{i<x,\theta>} e_1(t,x,\theta)\hat{f}(\theta)d\theta$$

$$E_2 f(t,x) = \int e^{i<t+x,\theta>} e_2(t,x,\theta)\hat{f}(\theta)d\theta$$

$$E_3 f(t,x) = \int_{-t}^{t} \int e^{i<\frac{r+t}{2} + x,\theta>} e_3(r,t,x,\theta)\hat{f}(\theta)d\theta dr$$

with

$$e_1, e_2 \in S^0(R^n \times R^{n-1})$$

$$e_3 \in S^1((R \times R^n) \times R^{n-1})$$

$$<x,\theta> = x_1\theta_1 + x_2\theta_2 + \cdots + x_n\theta_n$$

$$<t+x,\theta> = (t+x_1)\theta_1 + x_2\theta_2 + \cdots + x_n\theta_n$$

$$<\frac{r+t}{2} + x,\theta> = ((\frac{r+t}{2}) + x_1)\theta_1 + x_2\theta_2 + \cdots + x_n\theta_n$$

We need a further study of $E$ in order to get that $WF'E$ is concentrated along points where $A$ is a Fourier integral operator.

Lemma 5.1. Let $E$ be as in (3.1) a parametrix for the Cauchy problem for $K$. Then

$$WF'E \cap H = \phi$$

with

$$H = \{((t,x,\tau,\xi),(\bar{x},\bar{\xi})) \in T^*(R^n \times R^{n-1}) - \{0\} | t = 0,\ \xi = \bar{\xi} = 0\}.$$

Proof: Clearly $WF'E_i \cap H = \phi$, $i = 1,2$, since $E_1$(resp. $E_2$) is a Fourier integral operator associated with a parametrix for the Cauchy problem for

$$\left(\begin{pmatrix} D_t & 0 \\ 0 & D_t \end{pmatrix} + H(t,x,D_t)\right) \quad (\text{resp.} \quad \left(\begin{pmatrix} D_t - D_{x_1} & 0 \\ 0 \quad 0 & D_t - D_{x_1} \end{pmatrix}\right. + H(t,x,D_x)).$$

$K_{E_3}$ denote the Schwartz kernel of $E_3$, $f,\phi$ compactly supported $C^\infty$ functions. We have

$$K_{E_3}(\phi e^{-i<\cdot,\alpha(\tau,\xi)} \otimes f\, e^{-i<\cdot,\alpha\bar{\xi}>}) =$$

$$\int\int_{-t}^{t}\int \exp\{i(<\frac{r+t}{2} + x,\theta> - <(t,x),\ \alpha(\tau,\xi)>)\} \tag{5.2}$$

$$e_3(r,t,x,\theta)\hat{f}(\theta + \alpha\bar{\xi})\phi(t,x)d\phi drdtdx.$$

Also

$$\begin{pmatrix} D_t - D_r & 0 \\ 0 & D_t + D_r - D_{x_1} \end{pmatrix} e^{i(<\frac{r+t}{2} + x,\theta> - <(t,x),\alpha(\tau,\xi)>)} =$$

(5.3)

$$\begin{pmatrix} (\alpha\tau) & 0 \\ 0 & -\alpha(\tau-\xi_1) \end{pmatrix} \exp(i(\frac{r+t}{2} + x,\theta) - ((t,x),\alpha(\tau,\xi)).$$

Let us take $\phi$ supported near $(0,x_o)$ and $\phi = 1$ near $(0,x_o)$; $\xi$ near $0,\overline{\xi}$ near $0$. Repeated integration by parts in (5.2) using (5.3) and making the change of variables $\theta = \alpha\theta'$ gives that (5.1) = (5.4) with

$$= \alpha^{-(n-1)} \int\int_{-t}^{t}\int \exp i(<\frac{r+t}{2} + x,\theta> - <(t,x),\ \alpha(\tau,\xi)> \tag{5.4}$$

$$k_n\, e_3^{(n)}\ \hat{f}(\alpha(\theta' + \overline{\xi}))\psi_n d\theta' drdtdx$$

$$+ I_1 + I_2$$

where

$$k_n = \begin{pmatrix} -(\alpha\tau)^{-1} & 0 \\ 0 & -(\alpha(\tau-\xi_1))^{-1} \end{pmatrix}^n ,n \in N.$$

$e_3^{(n)}$ involves sum of derivatives of $e_3$ of order at most $n$ and $\psi_n$ involves derivatives up to order $n$ of $\phi$ and

$$I_1 = \int\!\!\cdot\!\!\int e^{i(<x,\theta> - <((t,x),\ \alpha(\tau,\xi)>)}$$

$$\tilde{e}_1(t,x,\theta)\hat{f}(\theta + \alpha\overline{\xi})k_n(t,x)d\theta drdtdx.$$

Similar expression for $I_2$ with $\tilde{e}_1$ replaced by $\tilde{e}_2$ and $<x,\theta>$ by $<t + x,\theta>$, $\tilde{e}_1,\tilde{e}_2 \in S^0(R^n \times R^{n-1})$, $\tilde{k}_n \in C_o^\infty(R^n)$ supported near $(0,x_o)$.

For $|\xi|$ small, $|\xi_1|$ is small; $|\overline{\xi}|$ is small, thus $\theta' + \overline{\xi} \approx \theta'$ and $\tau - \xi_1 \approx \tau$. Thus from (5.4) we get that

$$(5.4) = O(\alpha^{-N}) \ \forall \ N \in R.$$ Q.E.D.

Lemma 5.1 allows us to localize $E$ so that $WF'E$ is concentrated along points where $A$ as in (4.14) is Fourier integral operator.

This finishes the construction for $L_o$, thus we can construct a parametrix for the Cauchy problem for $P$ as in (1.1) satisfying (1.2) and (1.3).

Remark: The relationship with the phenomenon of "wedge refraction" is that the wave front set of $E_3$ is a wedge between the canonical relations generated by $WFE_1$, $WFE_2$ in the points of double characteristics. See [4] for more details.

## References

1. Hormander L. and Duistermaat J.J. Fourier Integral Operators II, Acta Math. 128, 183-269(1972).

2. Melrose, R.B., Transformation of boundary value problems, to appear in Acta Mathematica.

3. Melrose, R.B. and Uhlmann, G.A. Lagrangian intersection and the Cauchy problem, Comm . on Pure and Appl. Math. Vol. XXXII, 483-519(1979).

4. Uhlmann, G.A., Pseudodifferential operators with involutive double characteristics, Comm. in P.D.E., 2(7) 713-779(1977).

# 5
# A New Method of Computing Chromatic Polynomials of Graphs

ROBERTO W. FRUCHT* / Departamento de Matemáticas, Universidad Técnica Federico Santa María, Valparaíso, Chile

## 1. INTRODUCTION

The chromatic polynomials, first introduced by Birkhoff [2] in 1912, have their origin in the following consideration: A (point) *coloring* of a labeled graph G (where $V(G) = \{v_1,v_2,\ldots,v_p\}$ is the set of points of G) from $\lambda$ colors is an assignment of $\lambda$ *or fewer* colors to the points of G in such a way that no two adjacent points have the same color. Using the numbers 1, 2, 3, ..., $\lambda$ for the colors (instead of more pictorial names like blue, green, red, ...) we can also define a coloring as a function $f\colon V(G) \to \{1,2,\ldots,\lambda\}$ such that $f(v_i) \neq f(v_j)$ whenever $v_i$ and $v_j$ are adjacent.

Two such colorings of G from $\lambda$ colors will be considered different if at least one of the labeled points is assigned different colors. Let us then denote by $P(G,\lambda)$ the number of different colorings of a labeled graph G from $\lambda$ colors. It is well known that for any graph G with p points, $P(G,\lambda)$ is a polynomial of degree p in $\lambda$; it is called the *chromatic polynomial* (or *chromial*) of G.

For a proof and more information the reader might consult [7]; we only mention the following examples:

(i) The complete graph $K_p$ obviously has the chromatic polynomial

$$P(K_p,\lambda) = \lambda(\lambda - 1)(\lambda - 2)\cdots(\lambda - p + 1) = p!\binom{\lambda}{p} \tag{1.1}$$

which can also be written as

$$P(K_p,\lambda) = (\lambda)_p \tag{1.2}$$

Here for brevity we use the notation

$$(\lambda)_n = \lambda(\lambda - 1)(\lambda - 2)\cdots(\lambda - n + 1) \qquad 1 \le n \le \lambda \tag{1.3}$$

for "descending (or falling) factorials."

(ii) For the complement $\bar{K}_p$ we have trivially

$$P(\bar{K}_p,\lambda) = \lambda^p \tag{1.4}$$

(iii) The chromatic polynomial of a cycle or p-gon ($p \ge 3$) is

$$P(C_p,\lambda) = (\lambda - 1)\{(\lambda - 1)^{p-1} + (-1)^p\} \tag{1.5}$$

(See [7, Theorem 6].) In particular, for a square ($p = 4$) we have

$$P(C_4,\lambda) = \lambda(\lambda - 1)(\lambda^2 - 3\lambda + 3) \tag{1.6}$$

One can easily check, or see in [4], that the following are three other ways of writing the same polynomial:

$$P(C_4,\lambda) = \lambda^4 - 4\lambda^3 + 6\lambda^2 - 3\lambda \tag{1.7}$$

$$P(C_4,\lambda) = \lambda(\lambda - 1)^3 - \lambda(\lambda - 1)^2 + \lambda(\lambda - 1) \tag{1.8}$$

$$P(C_4,\lambda) = (\lambda)_4 + 2(\lambda)_3 + (\lambda)_2 \tag{1.9}$$

The same possibility of expanding the chromatic polynomial in different ways also exists, of course, for any other graph and has been widely used; the advantages and disadvantages of the different expansions are discussed in [6].

In this note we shall mainly be interested in expansions of the type (1.9), i.e., as a linear combination of the descending factorials defined by (1.3), say:

$$P(G,\lambda) = \sum_{i=1}^{p} k_i(G)(\lambda)_i \tag{1.10}$$

Because of (1.2) this so-called factorial expansion can also be written in the perhaps more pleasant form

$$P(G,\lambda) = \sum_{i=1}^{p} k_i(G)P(K_i,\lambda) \tag{1.11}$$

A reduction process for obtaining this kind of expansion for a given graph is well known; see [5, Fig. 12.15]. In this note a direct method for determining the coefficients $k_i(G)$ will be derived. Curiously enough, this method is implicitly contained in the reasoning leading to the well known combinatorial interpretation of these coefficients (see [7, Theorem 15]), but it seems that our method has not been stated explicitly in the literature.

Since it consists of counting the number of certain subgraphs in the complement $\bar{G}$, this counting process will be explained in Section 2 and then applied to our problem in Section 3.

## 2. SPECIAL SPANNING SUBGRAPHS

It will be convenient to have a short name for those *spanning* subgraphs S in a graph G—i.e., subgraphs with point set V(S) = V(G)—whose connected components are *complete* graphs (including $K_1$'s).

DEFINITION 2.1. A spanning subgraph will be called *special* whenever its connected components are complete graphs.

In other words, a special spanning subgraph of a graph with p points is a subgraph of the form

$$S = K_\alpha \cup K_\beta \cup K_\gamma \cup \cdots \tag{2.1}$$

where

$$\alpha + \beta + \gamma + \cdots + = p \tag{2.2}$$

In the next section we shall need certain numbers $s_i(G)$ defined as follows for $1 \le i \le p$ and for any labeled graph G:

$$s_i(G) = \text{number of special spanning subgraphs with } i \text{ connected components} \tag{2.3}$$

Of course, we have always

$$s_p(G) = 1 \tag{2.4}$$

and for any nontrivial graph with q lines:

$$s_{p-1}(G) = q \qquad (p \ge 2) \tag{2.5}$$

Indeed, there is only one spanning subgraph with p components, namely the totally disconnected subgraph $\bar{K}_p$ without lines, being the union of p $K_1$'s; and special spanning subgraphs with p - 1 components are necessarily of the form

$$K_2 \cup (p - 2)K_1$$

and so their number equals the number of lines in G, thus proving (2.5).

On the other hand, it might be remarked (although we shall not use this result later) that for any graphs G whose complement $\bar{G}$ has the chromatic number $\chi(\bar{G})$, we have

$$s_i(G) = 0 \qquad \text{for } i < \chi(\bar{G}) \tag{2.6}$$

This statement is equivalent to one to be found in [5, p. 129], namely that $\chi(\bar{G})$ is the minimum number of subsets which partition V(G) so that each subset induces a complete subgraph of G. Equation (2.6) also follows easily from (3.3) below.

EXAMPLE. $G = 2K_3$, the graph formed by two triangles; $p = q = 6$.

According to (2.4) and (2.5) we have

$$s_6(2K_3) = 1 \qquad s_5(2K_3) = 6 \tag{2.7}$$

Special spanning subgraphs with four components are either of the form $2K_2 \cup 2K_1$ or of the form $K_3 \cup 3K_1$. Now it is easily seen that in our example there are nine such subgraphs of the first kind and two of the second; hence:

$$s_4(2K_3) = 11 \tag{2.8}$$

It is easily checked that in our example the only special spanning subgraphs with three components are of the form $K_3 \cup K_2 \cup K_1$, and there are six of them; hence:

$$s_3(2K_3) = 6 \tag{2.9}$$

Finally, there is only one special spanning subgraph with two components, namely the graph itself, and none with one component:

$$s_2(2K_3) = 1 \qquad s_1(2K_3) = 0 \tag{2.10}$$

REMARKS. (i) Special spanning subgraphs with i components obviously correspond to partitions of the number p into i summands; e.g., the six subgraphs of the form $K_3 \cup K_2 \cup K_1$ in our example correspond to the partition $3 + 2 + 1 = 6$. The converse, however, is not always true; there can be partitions of the number p to which no special spanning subgraph belongs (e.g., in our example the partitions 2 + 2 + 2 and 4 + 1 + 1; there are indeed no spanning subgraphs of the forms $3K_2$ or $K_4 \cup 2K_1$ in $2K_3$).

(ii) *If the labeled graph* G *contains no triangles*, instead of the definition (2.3) the following, more simple formula can be used for $1 \le i < p$:

$$s_i(G) = \text{number of subgraphs of the form } (p - i)K_2 \tag{2.11}$$

In fact, in the absence of triangles (and hence of $K_n$'s in general for $n > 2$), the only special spanning subgraphs with i components that possibly can exist are of the form

$$(p - i)K_2 \cup (2i - p)K_1 \tag{2.12}$$

corresponding to the partition

$$p = \underbrace{2 + 2 + \cdots + 2}_{(p - i) \text{ summands}} + \underbrace{1 + 1 + \cdots + 1}_{(2i - p) \text{ summands}} \tag{2.13}$$

Disregarding in (2.12) the (2i - p) trivial components, we see that (2.11) follows immediately.

EXAMPLE. $G = C_6$ (the hexagon), $p = q = 6$.

From (2.4) and (2.5) we have

$$s_6(C_6) = 1 \qquad s_5(C_6) = 6 \tag{2.14}$$

then from (2.11)

$$s_4(C_6) = 9 \tag{2.15}$$

because the desired subgraphs of type $2K_2$ correspond in the hexagon either to a pair of opposite sides (three cases) or to a pair of neither opposite nor adjacent sides (six cases). And counting triples of mutually disjoint sides we find

$$s_3(C_6) = 2 \tag{2.16}$$

Finally,

$$s_1(C_6) = s_2(C_6) = 0 \tag{2.17}$$

In an analogous fashion it is possible to find a general formula for $s_i(C_p)$, namely,

$$s_i(C_p) = \frac{p}{i}\binom{i}{p-i} \qquad p \geq 4,\ 1 \leq i \leq p \tag{2.18}$$

The proof of (2.18) is left to the reader as an exercise.

## 3. COMPUTING CHROMATIC POLYNOMIALS BY COUNTING THE SPECIAL SPANNING SUBGRAPHS IN THE COMPLEMENTARY GRAPH

Why our interest in the numbers $s_i(G)$ introduced in the foregoing section? The answer is: it turns out that these numbers, when computed for the complement $\bar{G}$ rather than for the given graph G, are equal to the coefficients $k_i(G)$ in (1.10) or (1.11) which we want to determine:

$$k_i(G) = s_i(\bar{G}) \qquad i = 1, 2, \ldots, p \tag{3.1}$$

In fact, we shall prove in this section the following

THEOREM 3.1. Counting in $\bar{G}$ the number of special spanning subgraphs with i components, we obtain the coefficient $k_i(G)$ in the factorial expansion (1.10) or (1.11) of the chromatic polynomial $P(G,\lambda)$ $(i = 1, 2, \ldots, p)$.

PROOF. It is only a rephrasing of Read's Theorem 15 in [7, p. 65] and of the considerations leading to it, when we begin by stating that the coefficient $k_i(G)$ equals the number of possible partitions of V(G) into i mutually disjoint subsets, say $S_1, S_2, \ldots, S_i$, such that the induced subgraphs $\langle S_1\rangle, \langle S_2\rangle, \ldots, \langle S_i\rangle$ *have no lines.*

But then the corresponding subgraphs in the complement $\bar{G}$ are *complete* graphs, and their union is just what we have called a special spanning subgraph with i components. On the other hand, since each such subgraph in $\bar{G}$ obviously gives rise to a partition in G of the kind just described, the coefficient $k_i(G)$ is equal to $s_i(\bar{G})$ as had to be shown.

COROLLARY 3.1. For any graph G we have

$$P(G,\lambda) = \sum_{i=1}^{p} s_i(\bar{G})(\lambda)_i \tag{3.2}$$

and, of course, for the complement $\bar{G}$:

$$P(\bar{G},\lambda) = \sum_{i=1}^{p} s_i(G)(\lambda)_i \tag{3.3}$$

EXAMPLES. (i) The bicomplete graph $K_{3,3}$ ("4-cage") has two triangles as its complement; so it follows from (2.7)-(2.10) that:

$$P(K_{3,3},\lambda) = (\lambda)_6 + 6(\lambda)_5 + 11(\lambda)_4 + 6(\lambda)_3 + (\lambda)_2 \qquad (3.4)$$

(ii) The triangular prism $T_3$ (see Fig. 1) has a hexagon $C_6$ as its complement; so it follows from (2.14)-(2.17) that

$$P(T_3,\lambda) = (\lambda)_6 + 6(\lambda)_5 + 9(\lambda)_4 + 2(\lambda)_3 \qquad (3.5)$$

FIGURE 1. Triangular prism.

Formulas (3.4) and (3.5) can be found in [4].

(iii) For higher values of p the application of (3.2) can become quite cumbersome. This can already be seen from the example of the cube graph $Q_3$ ($p = 8$, $q = 12$) whose complement is shown in Figure 2.

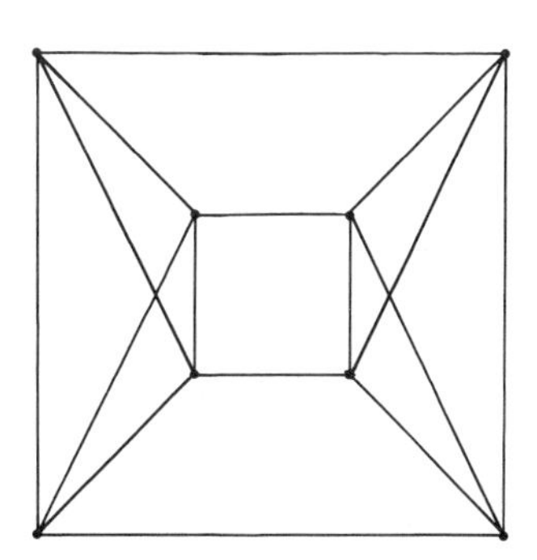

FIGURE 2. Complement of cube graph $Q_3$.

It actually took the author several hours to count all the special spanning subgraphs in Figure 2 without overlooking some, obtaining the result

$$P(Q_3,\lambda) = (\lambda)_8 + 16(\lambda)_7 + 80(\lambda)_6 + 146(\lambda)_5 + 92(\lambda)_4 + 18(\lambda)_3 + (\lambda)_2 \qquad (3.6)$$

(which was then checked by a computer program).

By the way, an explicit formula for the chromatic polynomial of a prism in general is to be found in [1, p. 63], but not in the factorial form that interests us here.

(iv) As an example of an application of formula (3.3) let us mention the possibility of obtaining the chromatic polynomial of the complement of a cycle (or p-gon) $C_p$ by using (2.18):

$$P(\bar{C}_p,\lambda) = \sum_{i=1}^{p} \frac{p}{i} \binom{i}{p-i} (\lambda)_i \qquad p \geq 4 \tag{3.7}$$

Essentially the same formula is already to be found in [6] as Theorem 5.1.3; Loerinc obtained it by a reduction process. In the same "report" she also gives similar formulas for several tree complements which can be checked by using our formula (3.3).

REMARKS. (i) In a subsequent paper, Theorem 3.1 will be exploited more systematically for finding new families of chromatically unique graphs. For a first step in this direction, see [3].

(ii) It is obvious and well known that the coefficients in the factorial expansion (1.10) are always nonnegative. Theorem 3.1 also allows us to find an upper bound for the same coefficients, namely,

$$k_i(G) \leq S(p,i) \qquad 1 \leq i \leq p \tag{3.8}$$

where the $S(p,i)$ are Stirling numbers of the second kind, defined by the identities

$$\lambda^p = \sum_{i=1}^{p} S(p,i)(\lambda)_i \tag{3.9}$$

In fact, for any graph G with p points, the number $s_i(\bar{G})$ of special spanning subgraphs with i components in the complement $\bar{G}$ is obviously at most equal to the same number $s_i(K_p)$ when $\bar{G}$ is a complete graph; but G is then the totally disconnected graph $\bar{K}_p$ with the chromatic polynomial (1.4) whose factorial expansion is given by (3.9). Hence,

$$s_i(\bar{G}) \leq S(p,i) \qquad 1 \leq i \leq p \tag{3.10}$$

and the inequality (3.8) now follows from (3.1).

## REFERENCES

1. N. E. Biggs, "Algebraic Graph Theory," Cambridge Tracts in Math., No. 67, Cambridge Univ. Press, London, 1974.

2. G. D. Birkhoff, "A determinant formula for the number of ways of coloring a map," Ann. Math. 14 (1912), 42-46.

3. R. E. Giudici, "Some new families of chromatically unique graphs," to in Analysis, Geometry, and Probability, edited by Rolando Chuaqui, Marcel Dekker, Inc., New York, New York, 1985, 147-159.

4. R. E. Giudici and R. M. Vinke, "A table of chromatic polynomials," J. of Combinatorics, Information and System Sciences 5 (1980), 323-350.

5. F. Harary, "Graph Theory," Addison-Wesley, Reading, Mass., 1969.

6. B. M. Loerinc, "Computing Chromatic Polynomials for Special Families of Graphs," Courant Computer Science Report No. 19 (Feb. 1980), Courant Institute of Math. Sciences, New York University.

7. R. C. Read, "An Introduction to Chromatic Polynomials," J. Combin. Theory 4 (1968), 52-71.

6

# On a Maximality Theorem in the Theory of Abstract Hardy Algebras Beyond the Szegö-Situation

LUIS SALINAS-CARRASCO / Departamento de Matemáticas, Universidad Técnica Federico Santa María, Valparaíso, Chile

## 1. INTRODUCTION

As usual in the theory of abstract Hardy algebras, we fix a positive measure space $(X,\Sigma,m)$ and consider the Banach algebra $L^\infty(m)$ equipped also with the weak-$*$ topology $\sigma(L^\infty(m),L^1(m))$, i.e., the coarsest topology of $L^\infty(m)$ such that all functionals $f \mapsto \int fFdm$ for all $f \in L^\infty(m)$ and $F \in L^1(m)$ are continuous on $L^\infty(m)$. The abstract Hardy algebra situation consists of a weak-$*$ closed complex subalgebra H of $L^\infty(m)$ containing the constants, together with a nonzero multiplicative linear functional $\varphi$: $H \to \mathbb{C}$ which is weak-$*$ continuous, i.e., $\varphi$ allows a representation in the form $\varphi(u) = \int uFdm \ \forall\ u \in H$ for some $F \in L^1(m)$. Then it is well known (cf. [1], Chap. IV) that $\varphi$ can be represented by nonnegative functions $0 \le F \in L^1(m)$. Our standard reference for notations, definitions, and known results will be [1]. Nevertheless, for the sake of completeness we recall the next few definitions: the set of all nonnegative representative densities $0 \le F \in L^1(m)$ is denoted by M, and the real linear span of the difference set M - M is denoted by N; furthermore, K denotes the set of all $f \in L^1(m)$ such that $\int ufdm = \varphi(u) \int fdm \ \forall\ u \in H$. Obviously M is a convex closed subset of the unit ball of the Banach space Re $L^1(m)$, and K is a closed complex subspace of $L^1(m)$. Clearly $H \subset K$ and $N \subset K$.

In 1972 P. S. Muhly [2] working in the so-called weak-$*$ Dirichlet or Szegö situation, which is characterized by the fact that Re H is weak-$*$ dense in Re $L^\infty(m)$ or, equivalently, by dim N = 0, showed the equivalence of the following properties which are of a quite different nature: (i) H has no zero divisors, (ii) for each nonzero $h \in H$ holds $m([h = 0]) = 0$, and (iii) H is a maximal proper weak-* closed complex subalgebra of $L^\infty(m)$. Muhly's proof was then simplified first in [3] and later in [1]. The following new equivalent properties were then added to the above list: (iv) if Y is an invariant weak-* closed subspace of $L^\infty(m)$ such that $fY \subseteq Y$ for $f \in L^\infty(m)$ with $f \notin H$, then $Y = \chi_E L^\infty(m)$ for some $E \in \Sigma$, and (iv') for all $E \in \Sigma$ the weak-* closure of $\chi_E H$ is $\chi_E L^\infty(m)$. It is noteworthy that in the classic disk situation all these properties hold true (cf. [1], I).

The main purpose of this note is to study the extension of the aforementioned equivalences to the particular weak-* hypo-Dirichlet situation where the weak-* closure of Re H has codimension 1 in Re $L^\infty(m)$ or, equivalently, dim N = 1 (cf. [1], VI).

## 2. PRELIMINARY RESULTS

Let us consider a reduced Hardy algebra situation (cf. [1], p. 64). Then the function class E of all *conjugable* functions is defined to consist of the functions $P \in \text{Re } L(m)$ such that there exists a function $Q \in \text{Re } L(m)$ such that $e^{t(P+iQ)} \in H^\#$ for all $t \in \mathbb{R}$ (cf. [1], p. 111), where the function class $H^\#$ consists of all $f \in L(m)$ such that there is a sequence $u_n \in H$ $(n = 1, 2, 3, \ldots)$ with $|u_n| \leq 1$, $u_n f \in H$, and $u_n \to 1$ pointwise in the L(m) sense as $n \to \infty$ (cf. [1], p. 69). It is not hard to see that E is a real vector space and $H^\#$ is an algebra. Furthermore, $E^\infty = E \cap \text{Re } L^\infty(m)$ satisfies $E^\infty = N^\perp$ (cf. [1], VI: 3.4, p. 116).

2.1 LEMMA. Let dim N = 1 and $f \in L^\infty(m)$. Then there is an $\alpha \in \mathbb{C}$, $\alpha \neq 0$, such that $\text{Re}(\alpha f) \in E^\infty$.

PROOF. Let $0 \neq G \in N$. Then $N = \mathbb{R}G \subset \text{Re } L^1(m)$, since dim N = 1. If $\int fGdm = 0$, then clearly $\text{Re } f \perp G$ and hence $\text{Re } f \in E^\infty$. If $\int fGdm = \beta \neq 0$, then $\int i\bar{\beta}fGdm = i|\beta|^2$ and hence $\int \text{Re}(i\bar{\beta}f)Gdm = 0$, so that $\text{Re}(i\bar{\beta}f) \in E^\infty$. ∎

Another result which will be useful in the sequel is the next general lemma whereof the proof is independent of the theory of Hardy algebras. We must introduce first the notion of the *transporter* $\tau(T)$ of a weak-* closed linear subspace $T \subseteq L^\infty(m)$; $\tau(T)$ is defined as the set of all $f \in L^\infty(m)$ such that $fT \subseteq T$ (cf. [1], p. 152). It is not hard to see that $\tau(T)$ is a weak-* closed complex subalgebra of $L^\infty(m)$ containing the constants.

2.2 LEMMA. Let $T \subseteq L^{\infty}(m)$ be a weak-* closed linear subspace. Then $\tau(T) = L^{\infty}(m)$ if and only if $T = \chi_V L^{\infty}(m)$ for some $V \in \Sigma$.

We omit the proof (cf. [1], p. 152).

## 3. MAIN RESULT

In the sequel we fix a reduced Hardy algebra $H \neq \mathbb{C}$. Our purpose is to prove Muhly's Maximality Theorem ([3], p. 17, and [1], p. 152) under the weaker assumption dim $N = 1$.

3.1 THEOREM. The subsequent properties are equivalent:

(i) $H$ is a maximal proper weak-* closed subalgebra of $L^{\infty}(m)$.

(ii) If $T \subseteq L^{\infty}(m)$ is an invariant weak-* closed linear subspace such that $fT \subseteq T$ for some $f \in L^{\infty}(m)$ with $f \notin H$, then $T = \chi_V L^{\infty}(m)$ for some $V \in \Sigma$.

(iii) For any $u \in H$ with $u \neq 0$, we have $m([u = 0]) = 0$.

(iv) $H$ contains no zero divisors.

Major portions of the proof of the above theorem given for the Szegö situation dim $N = 0$ in [3] and [1], also apply to the weak-* hypo-Dirichlet situation dim $N = 1$. Other steps deserve special attention. For the sake of completeness we give a detailed proof. The proof will be after the scheme (i) ⇒ (ii) ⇒ (iii) ⇒ (iv) ⇒ (i). By far the most difficult part is again (iv) ⇒ (i). The proof of (i) ⇒ (ii) is now standard and (iii) ⇒ (iv) is obvious.

PROOF. (i) ⇒ (ii): Since $T$ is a weak-* closed invariant subspace of $L^{\infty}(m)$, the transporter $\tau(T)$ of $T$ is a weak-* closed complex subalgebra of $L^{\infty}(m)$ containing $H$ but being $\neq H$ because of the existence of some $f \in L^{\infty}(m)$, $f \notin H$ with $f \in \tau(T)$. Hence $\tau(T) = L^{\infty}(m)$ after (i). Then $T = \chi_V L^{\infty}(m)$ for some $V \in \Sigma$ because of 2.2.

(ii) ⇒ (iii): Fix a nonzero $u \in H$ and assume $m([u = 0]) > 0$. Then it is not hard to see that $T = \{f \in L^{\infty}(m) \mid fu \in H\}$ is an invariant weak-* closed linear subspace of $L^{\infty}(m)$ containing $H$. Furthermore, $\chi_{[u=0]} L^{\infty}(m) \subseteq T$ and hence $\chi_{[u=0]} T \subseteq T$. Since $0 \neq \chi_{[u=0]} \notin H$ by assumption, we have $T = \chi_V L^{\infty}(m)$ for some $v \in \Sigma$ in view of (ii). Now, since $H \subseteq T$, we get $m(V) = m(X)$ and hence $T = L^{\infty}(m)$. But then, in particular $\bar{u} \in T$ or $|u|^2 \in H$ and hence $u$ = constant because of the reducedness of the Hardy algebra (cf. [1], p. 65). Therefore $m([u = 0]) > 0$ implies $u = 0$, which is a contradiction. ∎

The proof of (iv) ⇒ (i) requires the subsequent lemma, as in the case dim N = 0. But in the present case dim N = 1, the proof of this lemma is more difficult since now we do not have a factorization theorem (cf. [1], p. 83) as in the Szegö situation. We give a detailed proof.

3.2 LEMMA. Let B be a weak-$*$ closed complex subalgebra of $L^\infty(m)$ such that $H \subset B$ but $H \neq B$. Define $\Delta = \{D \in \Sigma | \chi_D \in B\}$. Then $\Delta$ is a $\sigma$-algebra $\subsetneq$ $\Sigma$ containing all sets of $\Sigma$ with 0- and full $m(X)$-measure and satisfying the following property: For each nonzero $f \in L(m)$ there exists a set $D \in \Sigma$ such that $f\chi_D \neq 0$ and $f\chi_{X\setminus D} \neq 0$.

PROOF. (0) Clearly $\Delta \subsetneq \Sigma$ and for all $D \in \Sigma$ with $m(D) = 0$ or $m(X)$ we have $\chi_D \in \Delta$ and hence $D \in \Delta$ for all such D's. That $\Delta$ is a $\sigma$-algebra is also obvious. In the next steps we will study $\Delta$, but we will first show that $\Delta$ is not trivial, i.e., does not reduce to sets of $\Sigma$ with measure 0 or $m(X)$, in step (5).

(1) We will say that an $f \in L(m)$ is *$\Delta$-measurable* iff there exists a $\Delta$-measurable function $\tilde{f}$: $X \to \mathbb{C}$ which is in the function class f. Then, clearly, every representative function $\tilde{f}$: $X \to \mathbb{C}$ of f is $\Delta$-measurable since all 0-measure sets of $\Sigma$ are in $\Delta$. Furthermore, we define $L(m|\Delta)$ as the set of all $f \in L(m)$ which are $\Delta$-measurable.

We will say that a set $D \in \Delta$ with $m(D) > 0$ is *minimal* iff for all $U \in \Delta$ with $U \subsetneq D$ (mod m) holds $m(U) = 0$ or $m(D\setminus U) = 0$. The main result of step (6) will be that $\Delta$ has no minimal set and this will allow us in step (7) to show the present lemma.

(2) We recall that for $f \in L(m|\Delta)$, for every representative function $\tilde{f}$: $X \to \mathbb{C}$ of f and for all open sets $S \cong \mathbb{C}$ the set $\tilde{f}^{-1}(S)$, i.e., $\{x \in X| \tilde{f}(x) \in S\}$, is $\Delta$-measurable or, equivalently, $\chi_{\tilde{f}^{-1}(S)} \in B$. In particular, for $\alpha \in \mathbb{C}$ and $\delta > 0$ the set:

$$[|f - \alpha| < \delta] = \{x \in X: \ |f(x) - \alpha| < \delta\} = \tilde{f}^{-1}(D(\alpha,\delta)) \tag{3.1}$$

is in $\Delta$.

(3) Now let $D \in \Delta$ be such that $m(D) > 0$ and D is minimal. Of course, we still do not know if such a D exists! However, we will get some conclusions about such a D, assuming its existence, which eventually will lead us to a contradiction. For $\alpha,\beta \in \mathbb{C}$ with $\alpha \neq \beta$ we will write E' = $[|f - \alpha| < (1/2)|\alpha - \beta|] \in \Delta$, $F' = [|f - \beta| < (1/2)|\alpha - \beta|] \in \Delta$, $E = D \cap E'$ and $F = D \cap F'$. Then obviously $\chi_E = \chi_{D\cap E'} = \chi_D\chi_{E'} \in B$ and analogously $\chi_F \in B$, so that $E,F \in \Delta$. Now, clearly $E \cap F = \emptyset$ and $E,F \subsetneq D$, so that if we

assume $m(E), m(F) > 0$, we get $0 < m(E), m(F) < m(D)$ which is a contradiction to the minimality of D. Hence E and F cannot both have positive measure or, equivalently, $\alpha$ and $\beta$ cannot both be in the value carrier $\omega(f|D)$ of $f|D$ (cf. [1], p. 102). Therefore $f|D$ = constant.

(4) We now claim that all real-valued functions $f \in B$ are $\Delta$-measurable. In fact, assume F is a fixed representative function of some real-valued $f \in B$, so that F is, in particular, bounded, say $|F(x)| \leq R$ for all $x \in X$ for some $R \in \mathbb{R}^+$. Define now $\Omega$ as the set of all functions $h\colon \mathbb{R} \to \mathbb{C}$ such that $h \circ F \pmod{m} \in B$. Then, clearly enough: (i) $\Omega$ contains all polynomials, (ii) for any given $h\colon \mathbb{R} \to \mathbb{C}$ such that there exists a sequence $h_n \in \Omega$ $(n = 1, 2, 3, \ldots)$ which is uniformly bounded on the interval $[-R,R] \subset \mathbb{R}$ and $h_n \to h$ pointwise on $[-R,R]$ for $n \to \infty$ (of course, we do not care about the behavior of h or the $h_n$'s on $\mathbb{R} \setminus [-R,R]$), then obviously $h \in \Omega$ since B is weak-$*$ closed, (iii) therefore, by a Stone-Weierstrass argument, $\Omega$ contains all continuous functions and hence all Baire functions $h\colon \mathbb{R} \to \mathbb{C}$ bounded on $[-R,R]$. Now, for any Baire set $A \subseteq \mathbb{R}$ we clearly have $\chi_A \circ F = \chi_{[F \in A]}$. Since, on the other hand, $\chi_A$ is a bounded Baire function on $\mathbb{R}$, by (iii) above we get $\chi_{[F \in A]} = \chi_A \circ F \pmod{m} \in B$ and therefore $[f \in A] \in \Delta$ as claimed.

(5) We now show that $\Delta$ is not trivial. More precisely, we will show that there exists $U \in \Delta$ such that $0 < m(U) < m(X)$. In fact, assume this does not hold or, equivalently, for all $E \in \Delta$ we have $m(E) = 0$ or $m(E) = m(X)$ or, again equivalently, X is minimal in the sense of (i). Then (3) and (4) imply $f = f|X$ = constant for all real-valued $f \in B$.

Fix now some arbitrary $f \in B$. By 2.1 there is an $\alpha \in \mathbb{C}$, $\alpha \neq 0$, such that $P = \mathrm{Re}(\alpha f) \in E^\infty$. Writing $Q = \mathrm{Im}(\alpha f)$ we then have: $\exp\{t(P + iP^*)\} \in H^X$ and $g_t = \exp\{t\alpha f\} \exp\{-t(P + iP^*)\} = \exp\{it(Q - P^*)\}$ for all $t \in \mathbb{R}$ (cf. [1], p. 111). Therefore $g_t \in B^X$, $|g_t| = 1$, $\bar{g}_t = 1/g_t \in B^X$ and hence $\mathrm{Re}(g_t), \mathrm{Im}(g_t) \in B$ for all $t \in \mathbb{R}$. By the above argument we then get that $\mathrm{Re}(g_t)$ = constant on $X = a(t)$ and $\mathrm{Im}(g_t)$ = constant on $X = b(t)$, i.e., $g_t$ = constant on $X = c(t)$ for all $t \in \mathbb{R}$.

For each $t \in \mathbb{R}$ consider now the function:

$$\exp(t\alpha f) = g_t \exp(t(P + iP^*)) = c(t) \exp(t(P + iP^*)) \in H^X \tag{3.2}$$

Then, clearly, for $t \in \mathbb{R}$ with $t \neq 0$ we have:

$$\frac{\exp(t\alpha f) - 1}{t} = f + \frac{(\alpha f)^2}{2!} t + \frac{(\alpha f)^3}{2!} t^2 + \cdots \in H \tag{3.3}$$

On the other hand, since for $|t| < 1$, $t \neq 0$, holds:

$$\left|\frac{\exp(t\alpha f) - 1}{t}\right| < 1 + |\alpha f| + \frac{1}{2!}|\alpha f|^2 + \frac{1}{2!}|\alpha f|^2 + \cdots = \exp(|\alpha f|) \quad (3.4)$$

i.e., $(\exp(t\alpha f) - 1)/t$ is uniformly bounded for $|t| < 1$ and $t \neq 0$. Therefore, the weak-* closedness of H implies that $\alpha f = \lim_{t\to 0}[(\exp(t\alpha f) - 1)/t] \in H$, i.e., $f \in H$. Since $f \in B$ was arbitrary, we conclude that $B \subseteq H$, which is a contradiction in view of $H \subset B$ and $H \neq B$. This contradiction shows that X is *not* minimal or, equivalently, that $\Delta$ is *not* trivial, as claimed.

(6) We now claim that $\Delta$ contains no minimal set in the sense of (1). In fact, fix some $D \in \Delta$. In view of (5) we may assume $0 < m(D) < m(X)$. Since $D \in \Delta$, clearly $\chi_D, \chi_{X\setminus D} \in B$ and hence $\exp(\chi_{X\setminus D}) \in B^X H$. Then there exists $\alpha \in H$, $\text{Re}\,\alpha > 0$, such that $\text{Re}\{\exp(\chi_{X\setminus D} + \alpha)\} \in E^\infty$. Now, it is not difficult to see that $\exp(n\chi_{X\setminus D}) = a_n \exp(\chi_{X\setminus D}) + b_n \in B^X/H$ for all $n \in \mathbb{N}$, where $a_n = (e^n - 1)/(e - 1) \in \mathbb{R}$ and $b_n = 1 - (e^n - 1)/(e - 1) \in \mathbb{R}$ $(n \in \mathbb{N})$. Therefore the same $\alpha \in H$, $\text{Re}\,\alpha > 0$, as before satisfies:

$$\text{Re}\{\exp(n\chi_{X\setminus D} + \alpha)\} = a_n \text{Re}\{\exp(\chi_{X\setminus D} + \alpha)\} + b_n \text{Re}\{\alpha\} \in E^\infty \quad (3.5)$$

since $\text{Re}\,H \subset E^\infty$. Let us write now $\exp(n\chi_{X\setminus D} + \alpha) = P_n + iQ_n$ with $P_n = \text{Re}\{\exp(n\chi_{X\setminus D} + \alpha)\} \in E^\infty$ and $Q_n \in \text{Re}\,L^\infty(m)$ for all $n \in \mathbb{N}$. Then, for each $n \in \mathbb{N}$ there is a $P_n^* \in \text{Re}\,L(m)$ such that $\exp(t(P_n + iP_n^*)) \in H^X$ for all $t \in \mathbb{R}$. Therefore, we have

$$b_{n,t} = \exp\{t \exp(n\chi_{X\setminus D} + \alpha)\}\exp\{-t(P_n + iP_n^*)\} = \exp\{it(Q_n - P_n^*)\} \quad (3.6)$$

i.e., $b_{n,t}$ is a unimodular function in $B^X$, for all $n \in \mathbb{N}$ and all $t \in \mathbb{R}$. In particular, we have $\exp\{-\exp(n\chi_{X\setminus D} + \alpha)\} = b_n f_n$, where $b_n = b_{n,-1} \in B^X$ is unimodular and $f_n = \exp\{-(P_n + iP_n^*)\} \in H^X$ for all $n \in \mathbb{N}$.

Now assume D is minimal in the sense of (1). Then as in (5), we get $b_n|D$ = constant. Since $|b_n| = 1$, without restriction we may assume $b_n|D = 1$. For any given $\varepsilon > 0$ and $n \in \mathbb{N}$ big enough $(n >> |\alpha|)$ we then have

$$\exp\{-\exp(n\chi_{X\setminus D} + \alpha)\} = \begin{cases} \eta_n & \text{on } X\setminus D \\ \exp\{-\exp\alpha\} & \text{on } D \end{cases} \quad (3.7)$$

where $\eta_n\colon X\setminus D \to \mathbb{C}$ satisfies $|\eta_n| < \varepsilon$ and $\eta_n \to 0$ pointwise boundedly on $X\setminus D$ for $n \uparrow \infty$. Since $|b_n| = 1$ we herefrom deduce:

$$fn = \begin{cases} \bar{b}_n\eta_n & \text{on } X\setminus D \\ \exp\{-\exp\alpha\} & \text{on } D \end{cases} \qquad (n \in \mathbb{N}) \quad (3.8)$$

Define now

$$f = \begin{cases} 0 & \text{on } X\setminus D \\ \exp\{-\exp \alpha\} & \text{on } D \end{cases} \tag{3.9}$$

Then clearly $f_n \to f$ pointwise boundedly on X for $n \uparrow \infty$, so that in view of the weak-* closedness of H, we get $f \in H$. But this is a contradiction since $f \in H$ implies $\chi_D \in H$ and $0 < m(D) < m(X)$. This proves (5).

(7) We are now in a position to finish the proof of the lemma. Fix some $0 \neq f \in L(m)$ and consider the set $\Delta(f) = \{D \in \Delta | f\chi_D = f\}$. A number of immediate consequences follow: (i) $\Delta(f) \subseteq \Delta$, (ii) $\Delta(f) \neq \emptyset$ since $X \in \Delta(f)$, (iii) for all $D \in \Delta(f)$ holds $m(D) > 0$, (iv) $\Delta(f)$ is closed under unions and finite intersections, (v) if $D \in \Sigma$ and $D_n$ (n = 1, 2, 3, ...) is a sequence in $\Delta(F)$ with $D_n \downarrow D$, then $D \in \Delta(f)$, (vi) for all $D \in \Delta(f)$ we have $m(D) \geqq m([f \neq 0]) > 0$.

Now define $\alpha = \operatorname{Inf}\{m(D) | D \in \Delta(f)\}$. Clearly $0 < m([f \neq 0]) \leqq \alpha \leqq 1$. Take a sequence $D_n$ (n = 1, 2, 3, ...) in $\Delta(f)$ with $m(D_n) \downarrow \alpha$ for $n \uparrow \infty$. By (iv) above we may assume $D_n \supseteq D_{n+1}$ (n = 1, 2, 3, ...). Taking $D = \bigcap_{n=1}^{\infty} D_n$ we get $D_n \downarrow D$ and $\chi_{D_n} \to \chi_D$ pointwise boundedly for $n \uparrow \infty$. Since by definition $\chi_{D_n} \in B$ and B is weak-* closed, we get $\chi_D \in B$, i.e., $D \in \Delta$, and obviously $\chi_D f = f$. Hence $D \in \Delta(f)$ and $m(D) = \alpha$. By (6) above, there is then a set $U \subset \Delta$ such that $U \subsetneq D$ and $0 < m(U) < m(D)$. Now assume the lemma is false or, equivalently, that for $V \in \Delta$ holds either $f\chi_V = f$ or $f\chi_V = 0$. But both situations are impossible for the above $U \in \Delta$, because they contradict the definition of D. This proves the lemma. ∎

PROOF OF (iv) ⇒ (i). Assume (i) does not hold, i.e., that there exists a weak-* closed algebra $B \subset L^{\infty}(m)$ with $H \subset B$, $H \neq B$, and $B \neq L^{\infty}(m)$. Since $\dim N < \infty$, there exists internal representative densities in M (cf. [1], p. 138). Fix an internal $G \in M$ and consider the weak-* closed subspace $S = B + (N/G) + i(N/G) = B + \mathbb{C}(U/G)$ where $N = \mathbb{R}U$ for some arbitrary $U \in N$

CASE I. $S \neq L^{\infty}(m)$. Define $T = \{f \in L^{\infty}(m) | fG \perp S = B + \mathbb{C}(N/G)\}$. A number of consequences follows from the definition.

First of all, T contains functions $\neq 0$. In fact, take any $0 \neq h \in L^1(m)$ such that $fG \perp S$. Then clearly $h \in K = (H^{\#} \cap L^1(Gm))G + \mathbb{C}N \subset L^1(m)$ (cf. [1], p. 144). Therefore $h = uG + \lambda U$ for some $u \in H^{\#} \cap L^1(Gm)$ and $\lambda \in \mathbb{C}$. Suppose $\lambda \neq 0$; then $\int h(U/G)dm = \int uUdm + \lambda \int (U^2/G)dm$. Since $h \perp (U/G)$, the first integral is = 0; and since $u \in H^{\#} \cap L^1(Gm)$ and $U \in N$, the second integral is also = 0. Hence $\int (U^2/G)dm = 0$; but this implies

$U = 0$, which is a contradiction. Thus $\lambda = 0$ and hence $h = uG$ or, equivalently, $h/G = u \in H^{\#} \cap L^1(Gm)$. Since $h/G \in H^{\#}$, there exists a sequence $u_n \in H$ $(n = 1, 2, 3, \ldots)$ such that $|u_n| \leq 1$, $u_n(h/G) \in H$ for all $n = 1, 2, 3, \ldots$ and $u_n \to 1$ in the $L(m)$-sense as $n \uparrow \infty$. Now consider for $b \in B$, $\lambda \in \mathbb{C}$, and $n \in \mathbb{N}$ the equation

$$\int \left(b + \lambda \frac{U}{G}\right) u_n \frac{h}{G}\, Gdm = \int (bu_n) hdm + \lambda \int \left(u_n \frac{h}{G}\right) Udm = 0 \tag{3.10}$$

Herefrom, for $f_n = u_n(h/G) \in H$ we deduce $f_n G \perp B + \mathbb{C}N = S$ and accordingly $f_n \in T$ for all $n \in \mathbb{N}$. Since $u_n \to 1$ as $n \to \infty$ we eventually get lots of nonzero functions in T.

Furthermore, $BT \subseteq T$, which is obvious since B is an algebra, and $T \subset H_\varphi \subset H$. In fact, for any $f \in T$ we conclude as above that $fG = vG$ for some $v \in H^{\#} \cap L^1(Gm)$ or, equivalently, $f = v \in H$ in view of $G > 0$ (cf. [1], p. 70). Now, since $fG \perp B \ni 1$ we get $\varphi(v) = \int vGdm = \int 1 \cdot fGdm = 0$, i.e., $v \in H_\varphi \subset H$.

We note also that T is a weak-* closed B-invariant subalgebra of H although we will not need this in the sequel. In fact, that T is a B-invariant subalgebra of H is already clear from above. The weak-* closedness of T follows immediately from the fact that T is actually the polar in $L^\infty(m)$ of the subspace $GB + \mathbb{C}N \subseteq L^1(m)$ (cf. [1], p. 125).

We may now prove (iv) ⇒ (i) in Case I. For any $0 \neq f \in T$ we get by Lemma 3.2 sets $D \in \Delta$, with $\chi_D, \chi_{X\setminus D} \in B$, of course, such that $0 \neq \chi_D f$, $\chi_{X\setminus D} f \in BT \subset T \subset H_\varphi \subset H$. This provides us with lots of zero divisors in H, which is in fact a contradiction to (i).

CASE II. $S = L^\infty(m)$. We note that B and $L^\infty(m)$ are both weak-* closed H-invariant subspaces of $L^\infty(m)$ and $\dim(L^\infty(m)/B) = 1 < \infty$. Through an application of a commutative operator algebra result for finite dimensional vector spaces, discussed in detail in [3], p. 63, and [1], p. 165, we get a $0 \neq P \in L^\infty(m)$ with $P \notin B$ and a $\neq 0$ weak-* continuous multiplicative linear functional $\vartheta$: $H \to \mathbb{C}$ such that $(u - \vartheta(u))P \in B$ for all $u \in H$. Without restriction we may assume $P = U/G$ (recall that $N = \mathbb{C}U$).

Now take a $0 \neq h \in L^1(m)$ such that $h \perp B$. Then clearly $h(u - \vartheta(u)) \perp B$ for all $u \in H$. In view of $(u - \vartheta(u))U/G \in B$ $(u \in H)$ we get $\int h(u - \vartheta(u))(U/G)dm = 0$ for all $u \in H$. Therefore we have $h(u - \vartheta(u)) \perp B + \mathbb{C}(U/G) = S = L^\infty(m)$ and hence $h(u - \vartheta(u)) = 0$ for all $u \in H$. This means that for every $u \in H$ the function $u - \vartheta(u) \in H$ actually lives on $X\setminus[h \neq 0] \in \Sigma$ where $0 < m([h \neq 0]) \leq m(X)$. Since H does not consist of constant functions only, we have in fact $0 < m([h \neq 0]) < m(X)$.

Since $0 \neq h \perp B$ we conclude as in Case I that $h \in K = H^{\#} \cap L^1(Gm)G + \mathbb{C}U$, i.e., $h = u_0 G + \alpha_0 U$ for some $u_0 \in H^{\#} \cap L^1(Gm)$ and $\alpha_0 \in \mathbb{C}$.

Assume that $\alpha_0 \neq 0$. Then consider $h/G = u_0 + \alpha_0(U/G) \in L^1(Gm)$. In view of $h(u - \vartheta(u)) = 0$ for $u \in H$ and since $G > 0$ we get

$$u_0(u - \vartheta(u)) = -\alpha_0 \frac{U}{G}(u - \vartheta(u)) \qquad \forall\, u \in H \tag{3.11}$$

Since $\alpha_0 \neq 0$, $U \neq 0$, and $\mathbb{C} \neq H$, we have $u_0 \neq 0$. Furthermore, $u_0 \neq$ constant. In fact $u_0 = \vartheta(u_0)$ implies $h = \vartheta(u_0)G + \alpha_0 U$ and hence $0 = \int hdm = \vartheta(u_0) \int Gdm = \vartheta(u_0)$. Therefore $m([u_0 \neq \vartheta(u_0)]) > 0$ so that in view of $h(u - \vartheta(u)) = 0$ for all $u \in H$, we must conclude

$$0 < m([u_0 = \vartheta(u_0)]), \qquad m([u_0 \neq \vartheta(u_0)]) < m(X) \tag{3.12}$$

From (3.11) we have in particular

$$\frac{u_0}{\alpha_0} = \begin{cases} -\dfrac{U}{G} & \text{i.e., real-valued, on } [u_0 \neq \vartheta(u_0)] \\[2ex] \dfrac{\vartheta(u_0)}{\alpha_0} & \text{i.e., constant, on } [u_0 = \vartheta(u_0)] \end{cases} \tag{3.13}$$

Since $u_0/\alpha_0 \in H^{\#} \cap L^1(Gm)$, from (3.13) we get:

$$0 = \int \frac{u_0}{\alpha_0} Udm = -\int_{[u_0 \neq \vartheta(u_0)]} \frac{U^2}{G} dm + \frac{\vartheta(u_0)}{\alpha_0} \int_{[u_0 = \vartheta(u_0)]} Udm \tag{3.14}$$

Since $G > 0$ and $U \neq 0$ is real-valued, (3.14) implies $0 \neq \vartheta(u_0)/\alpha_0 \in \mathbb{R}$ and $\int_{[u_0=\vartheta(u_0)]} Udm \neq 0$. Therefore

$$v_0 = \frac{u_0}{\alpha_0} - \frac{\vartheta(u_0)}{\alpha_0} = \begin{cases} -\dfrac{U}{G} - \dfrac{\vartheta(u_0)}{\alpha_0} & \text{i.e., real-valued, on } [u_0 \neq \vartheta(u_0)] \\[2ex] 0 & \text{on } [u_0 = \vartheta(u_0] \end{cases} \tag{3.15}$$

is a real-valued function in H, which in view of (3.12) is not constant on X. This is however a contradiction since the only real-valued functions in H are the constants. Hence $\alpha_0 \neq 0$ cannot hold.

Therefore we must assume $\alpha_0 = 0$ and hence $h/G = u_0 \in H^{\#}$, so that $u_0(u - \vartheta(u)) = 0$ for all $u \in H$. Choosing a sequence $u_n \in H$ with $|u_n| \leqq 1$ and $u_n u_0 \in H$ $(n = 1, 2, 3, \ldots)$ and $u_n \to 1$ pointwise in the L(m)-sense as $n \to \infty$ we conclude:

$$(u_n u_0)(u - \vartheta(u)) = 0 \qquad \forall\, u \in H \tag{3.16}$$

which eventually furnishes lots of zero divisors in H. This is a contradiction to property (i). Thus the proof of the main theorem is finished. ∎

## REFERENCES

[1] Barbey, K., Konig, H. Abstract Analytic Function Theory and Hardy Algebras, Springer-Verlag, Berlin, 1977.

[2] Muhly, P. S. Maximal weak-* Dirichlet algebras, Proc. Amer. Math. Soc. 36 (1972), 515-518.

[3] Salinas-Carrasco, L. Maximalitätssatze in der abstrakten Hardy-Räume Theorie, Dissertation, Universität des Saarlandes, Saarbrücken (West Germany), 1976.

# 7
# Models for Probability

ROLANDO CHUAQUI / Departamento de Matematica, Pontificia Universidad
Católica de Chile, Santiago, Chile

## INTRODUCTION

When analyzing a physical situation mathematically, the first step is to set up a mathematical model of it. This means, to replace the physical objects, relations between them, properties, etc. by abstract objects that, so to say, mirror them. That is, to form an abstract world that reflects the physical world.

What do we need for such an abstract world?. In the first place, the set of material objects that conform the physical world, are replaced by an abstract set $A$, which should be non empty. Then, the binary relations between these objects, by abstract set-theoretical relations $R$ i.e. $R \subseteq {}^2A$. Properties, by subsets of $A$, etc. Thus, an abstract world may be thought of as a relational system of the form $\langle A, R_i, a_j \rangle_{i \in I, j \in J}$ where the $R_i$'s are unary, binary, ternary or in general, n-ary relations over $A$ (i.e. subsets of ${}^nA$) and the $a_j$'s belong to $A$. The $a_j$'s represent distinguished elements of $A$, i.e. things we are especially interested in and that we can identify.

---

The work of the author was partially supported by the Organization of American States through its Regional Scientific and Technological Program (grant 563479) the Fundação de Amparo a Pesquisa do Estado de São Paulo (FAPESP), and the Dirección de Investigación (DIUC) of the Catholic University of Chile, (grant 93/81).

I would like to thanks Professors Newton C.A. da Costa, Irene Mikenberg, Leopoldo Bertossi and William N. Reinhardt for many useful comments and inspiring conversations.

I do not believe that it is possible to describe completely a physical situation; thus, any abstract world is just an approximation to physical reality.

The same method can be used to model economic, social or other types of situations. The objects involved in these may be other than physical, but the abstract systems are similar to those explained above.

Notice two limitations of the abstract structures considered up to now as abstract worlds. On the one hand, they are completely determined. Thus, for instance, any pair of objects $\langle a,b \rangle$ is, or is not, in the binary relation $R_i$. However, in actual fact is often not completely determined whether two objects are in given relation or not. In order to take care of this problem, I have developed what I call simple probability structures. On the other hand, relational systems represent an instantaneous cross section of a world that is really changing. It is not easy to formalize by relational systems causal relations. This second problem is dealt with my causal structures. A combination of simple probability and causal structures, gives the compound probability structures.

In this paper, which is mainly expository, I shall concentrate on simple probability structures, summarizing what has been obtained up to now in Chuaqui 1977a and 1982 and adding some improvements. The causal and compound probability structures, of Chuaqui 1980 and 1982, will not be discussed.

In Section 1, I will expound on the notion of abstract world in its 'dual' relation to the physical world and to a language. Section 2 deals with simple probability structures, also in their relations with the physical world and a fixed language.

Section 3 defines and gives several examples of simple probability structures. Sections 4 and 5 define the $\sigma$-field of events and a group of transformations that is needed for obtaining the probability measure. Section 4 contains the only novelty in this paper: the definition of the $\sigma$-field of events based on the simple probability structure. Section 6 introduces an $\mathcal{L}_{\omega_1\omega}$ formal language and defines a probability function on its sentences. It can be skipped by those not interested in the logical aspects. Section 7 relates some mathematical problems arising from this approach. Some are solved and many are open. Finally, Section 8 indicates possible applications to statistical inference and physics.

## 1. RELATIONAL STRUCTURES AS MODELS OF REALITY AND AS POSSIBLE MODELS OF A LANGUAGE

In the Introduction, we saw how a relational system $\mathfrak{A} = \langle A, R_i, a_j \rangle_{i\in I,\ j\in J}$ could be considered as an abstract model of nature. I believe that it is impossible to describe completely a physical situation. Thus, $\mathfrak{A}$ can only be considered as an approximation to the situation. It is an approximation in two senses. In the first place, as an abstract object it is an idealization of the situation involved. In the second place, many features of nature are disregarded, since it would be impossible to consider them all. Only what is thought relevant is taken into account.

Let us take an example. Suppose the situation involved is the throwing of a die. The universe $A$ of our relational system will contain the objects we want to talk about. In this case, $A$ consists of the faces of the die. Thus, we disregard many other features of the die. The only property we need in this case, is the property of coming up on the throw, let us call it $C$. $C$ is a one element of subset of $A$. We may also want to speak individually about each face and, hence, have all of them as distinguished elements. Thus, our relational structure will be $\langle A, C, 1,2,3,4,5,6 \rangle$.

There is another way of looking at our relation with nature. We use a language in order to describe physical reality. In this language, we mention objects and speak about their relations. When our statements match the physical reality, we say they are true. Based on truth, we can describe logical truth and the relation of logical consequence.

We have 'introduced' the physical world into mathematics. The next step is to mathematize the language. A natural language is approximated mathematically by mathematical structures, called formal languages. These formal languages are approximations to natural languages in the same way as relational systems approximate the actual world. In formal languages, some features of natural languages are disregarded and others are idealized. Anyway, the notion of sentence is defined mathematically.

Words in a natural language such as English are interpreted as referring to objects in the physical world. We also understand what it means for a proposition expressed in English to be true or false. Loosely speaking, an English proposition is true when it 'corresponds' to reality. Or, as Aristotle puts it,

"To say of what is that it is not, or of what is not that it is, is false, while to say of what is that it is, and of what is not that it is not, is true". (Metaphysics, Γ, 1011 b 26).

(See Tarski 1969, for a discussion of this conception of truth.)

The interpretation of formal languages is in relational systems, the abstract representation of possible worlds. Thus, a symbol may represent a relation or an object in the universe of the structure. For $\langle A,C,1,\ldots,6\rangle$ we should have a language with a symbol for $C$ and symbols for $1,\ldots,6$.

Although it is impossible to formulate a correct formal definition of truth for natural languages, this is possible for formal languages. This definition was first given by Tarski in 1931 (see Tarski 1935). Since both the formal languages and their interpretations (i.e. relational systems) are mathematical objects, the definition of truth is also mathematical. What we define is a relation between a sentence $\phi$ and a relational system $\mathcal{A}$, i.e. the relation of $\phi$ being true in $\mathcal{A}$ (or according to the interpretation $\mathcal{A}$). This relation is usually expressed symbolically by $\mathcal{A} \models \phi$, and is also read $\mathcal{A}$ is a model of $\phi$.

This formal definition of truth is now a basic part of logic and is taught in the first courses on the subject. For instance, it is explained in Mates 1964 and Enderton 1972. Also, from the definition of truth, that of logical consequence is easily obtained.

Thus we have seen the dual role of relational systems. On one hand, they are models of the world. On the other hand, they are models of sentences of a language. Notice that the word "model" has shifted in meaning. In the first case, the system is a model of the world, because the relations of the system represent faithfully the physical relations which are intended as their interpretations. Thus, in this case, the system is true to nature. In the second case, on the other hand, the sentences are interpreted in relational systems, the possible models of the language; a sentence $\phi$ may be true or false in a given relational system $\mathcal{A}$. If it is true, we say that the relational system $\mathcal{A}$ is a model of $\phi$.

This dual role will be preserved for simple and compound probability structures. They will serve as models of reality and interpretations of languages.

I shall end this section with some technical remarks on relational systems. *Relational systems* will be for us systems of the form $\mathcal{A} = \langle A, R_i \rangle_{i \in I}$ where A is a nonempty set and $R_i \subseteq {}^nA$ for a certain $n < \infty$. (${}^nA$ is the set of n-tuples of A; we identify ${}^1A$ with A.) The *similarity type* of $\mathcal{A}$ contains the index set $I$ and a function $\nu : I \to \omega$ ($\omega$ is the set of natural numbers) such that if $R_i \subseteq {}^nA$, then $\nu(i) = n$, i.e., $\nu(i)$ gives the arity of $R_i$. Thus, the similarity type is a pair $\langle I, \nu \rangle$. If we want to talk of the index set of $\mathcal{A}$, we shall write $I^{\mathcal{A}}$. Similarly, the $i^{th}$ relation in $\mathcal{A}$, is written $R_i^{\mathcal{A}}$. If $J \subseteq I^{\mathcal{A}}$, we shall write $\mathcal{A} \restriction J$ for $\langle A, R_i^{\mathcal{A}} \rangle_{i \in J}$.

In general, an appropriate system may be of the form $\langle A, R_i, O_j, a_k \rangle_{i \in I, j \in J, k \in K}$, where the $R_i$'s are relations, the $O_j$'s are operations (i.e. n-ary functions from A to A), and the $a_k \in A$. However, for simplicity's sake, I will only consider relations in the official presentation, since n-ary operations are also (n + 1)-ary relations, and an element $a_k$ of A can be replaced by the unary relation $\{a_k\}$. In the informal remarks and examples, however, operations and distinguished elements will appear, it being understood that they should be replaced by the corresponding relation.

## 2. SIMPLE PROBABILITY STRUCTURES AS MODELS OF REALITY AND AS POSSIBLE MODELS OF A LANGUAGE

Simple probability structures were first introduced in Chuaqui 1977a, although some ideas for them were already present in Chuaqui 1965 and a previous attempt to define them is contained in Chuaqui 1975. These structures were introduced as means of clarifying the nature of probability and the foundations of statistical inference. In this paper, however, I shall concentrate on the mathematical aspects of the theory; those interested in the Philosophy of Probability can consult my earlier papers, especially Chuaqui 1977a, 1979, 1980 and 1982.

What we would like is a structure $K$ that determines a Boolean algebra of events $B$ and a probability measure $\mu$, at least in most cases where probability has been applied. $K$ should embody the description of the situation in question that we accept as valid. In Chuaqui 1977a, I obtained $B$ and $\mu$ using $K$ and a formal language $\mathcal{L}$. To make $B$ and $\mu$ dependent on a language is not very natural, if we want to consider the probability structures as models of reality. Hence, in Chuaqui 1982, this dependence on a language was abandones, at the cost, however, of having $K$ and $B$ dedetermined by the situation, when just $K$ should be enough. In the present paper, I shall give a definition of $B$ in terms of $K$, thus attaining our aim. However, a word of caution is in order. Although it is true that given any simple probability structure $K$ a Boolean algebra ($\sigma$-complete) of events is determined, it may be the case that no measure or more than one measure satisfy the conditions that will be given below. When either of these happen, I would say that there is no adequate probability model of the situation involved. However, I have always been able to construct simple probability structures that determine a unique measure for all real situations for which these structures are intended.

The general plan is the following. The simple probability structure $K$ determines the Boolean algebra of events $B$ . $B$ is a $\sigma$-field of sets. $K$ and $B$, in their turn , determine a group $G^*$ of transformations of $B$. The measure $\mu$, will be a measure invariant under $G^*$. What is actually determined by $K$ is $B$ and $G^*$. There may be no measure or more than one measure $\mu$ invariant under $G^*$. When there is exactly one measure $\mu$ satisfying this condition, I would say that the simple probability structure determines an adequate probability measure.

Thus, the triple $\langle K, B, \mu\rangle$ is what constitutes a simple probability model. Let us see, now, its relation to a language. The simple probability structure $K$

has as its main constituent *the set of possible outcomes* $\mathbf{K}$. Since each possible outcome is mirrored by a relational system (or, in other words, by a model of a possible real situation), $\mathbf{K}$ is a set of relational systems (with a common universe and a common similarity type). $B$ is a $\sigma$-field of subsets of $\mathbf{K}$.

Let $\mathcal{L}$ be an appropriate formal language, i.e. $\mathcal{L}$ should have symbols for all relations in $\mathbf{K}$. We already have a definition of truth for each particular relational system $\mathfrak{A}$ in $\mathbf{K}$, and any sentence $\phi$ in $\mathcal{L}$. We can say that $\phi$ is true in $\mathbf{K}$ (in symbols, $\mathbf{K} \models \phi$), if $\phi$ is true in every element of $\mathbf{K}$. Similarly, $\phi$ is false in $\mathbf{K}$ if $\phi$ is false in every element of $\mathbf{K}$. The set $Mod_{\mathbf{K}}(\phi) = \{\mathfrak{A} : \mathfrak{A} \in \mathbf{K}, \mathfrak{A} \models \phi\}$ measures, in a way, the degree to which $\phi$ is true in $\mathbf{K}$. $Mod_{\mathbf{K}}(\phi) = \mathbf{K}$, means that $\phi$ is true in $\mathbf{K}$, and $Mod_{\mathbf{K}}(\phi) = \emptyset$, means that $\phi$ is false in $\mathbf{K}$. Thus, the measure $\mu$ should be defined on all sets $Mod_{\mathbf{K}}(\phi)$ for $\phi$ a sentence of $\mathcal{L}$, i.e. these sets should all be in $B$. Thus, $\langle K, B, \mu \rangle$ is a possible simple probability model for the language $\mathcal{L}$, if $\mathcal{L}$ has symbols for all relations in $\mathbf{K}$ and $\{Mod_{\mathbf{K}}(\phi)\colon \phi \text{ a sentence of } \mathcal{L}\} \subseteq B$.

The probability of $\phi$ under the interpretation $K$, is given by $P_K(\phi) = \mu(Mod_{\mathbf{K}}(\phi))$.

## 3. SIMPLE PROBABILITY STRUCTURES, EXAMPLES AND DEFINITION

I shall begin by giving some examples, which are modifications of those presented in my earlier papers.

Example 1. The choosing of a sample $S$ of $m$ elements from a finite population of balls $P$.

When we say '$S$ has $n$ red balls', we mean that one of the properties of the outcome was that the sample had $n$ red balls. As was said in Section 1, an ideal approximation of an outcome is a relational system that is a model of the outcome. This relational system should include as relations all those that we consider identifiable in principle, and relevant. In the case in question, these relational systems should be of the form

$$\mathfrak{A}_S = \langle P, R_0, \ldots, R_{m-1}, S, a \rangle_{a \in P}\,.$$

$P$ is the finite set of balls, $R_0, \ldots, R_{m-1}$ are fixed subsets of $P$ that represent the identifiable and relevant properties (for instance, red); $S$ is any subset of $P$ of $m$ elements (the particular sample corresponding to the outcome). We have distinguished every element of $P$, since we consider the individual balls as identifiable and, of course, relevant. For each subset $S$ of $m$ members there is a corresponding system $\mathfrak{A}_S$ ; $\mathbf{K}_0$, the set of possible outcomes, consists of all systems $\mathfrak{A}_S$ of the form described above.

Let us examine a possible outcome $\mathfrak{A}_S = \langle P, R_0, \ldots, R_{m-1}, S, a \rangle_{a \in P}$. The structure of the experiment is given by the set of all $\langle P,S \rangle$, for the $\mathfrak{A}_S$ in $\mathbf{K}_0$. $\langle P,S \rangle$ is called the *structural part of* $\mathfrak{A}_S$, denoted by $\mathfrak{A}_{S,\, st}$. Notice that $\langle P,S \rangle \neq \langle P,S' \rangle$ for $\mathfrak{A}_S \neq \mathfrak{A}_{S'}$. The other relations and distinguished elements in $\mathfrak{A}_S$ (namely, $R_0, \ldots, R_{m-1}$, and $a$, for each $a \in P$)have a different role. They do not describe the experiment itself, but serve to describe the events and, hence, to determine the algebra of events $\mathcal{B}$ (see Section 4). A way of characterizing these relations is the following. The properties $R_0, \ldots, R_{m-1}$ and, of course, the distingued elements $a \in P$ are intrinsic to the balls in $P$. That is, when we move the balls around or choose a sample, the balls do not change with respect to these properties. Also, these properties and distinguished elements are the same in all $\mathfrak{A}_S$. We may thus call $\langle P, R_0, \ldots, R_{m-1}, a \rangle_{a \in P}$ the *intrinsic part of* $\mathbf{K}_0$.

We may have different experiments performed on the same set of balls $P$. The probability structures for these different experiments, might have the same intrinsic part but different structural parts. For instance consider the experiment performed on the same set of balls $P$, with the same properties $R_0, \ldots, R_{m-1}$, that consists of the obtaining of a sample $Q$ containing exactly two balls with the property $R_0$. The intrinsic part of this set of possible outcomes, will be the same $\langle P, R_0, \ldots, R_{m-1}, a \rangle_{a \in P}$, but the structural parts will be $\langle P,Q \rangle$, with $Q$ any subset of $P$ containing two $R_0$-balls.

With the same set of balls $P$ and the same experiment $S$ of selecting a sample of size $m$, we also may have different intrinsic parts. For instance, we might decide to consider just one property, say $R_0$ (which may mean 'red'). Thus, the intrinsic part of this new experiment would be $\langle P, R_0, a \rangle_{a \in P}$ and the structure of the experiment, as before, $\langle P,S \rangle$. As will be seen in Section 4, this change in intrinsic part produces a change in the algebra of events.

Although in these examples we can distinguish the intrinsic and structural parts by looking at which relations are fixed in all $\mathfrak{A}_S$, and which are variable, in general, this is not so. I believe that the situation itself gives the structure of the experiment and the intrinsic part. Thus, a complete description of the phenomenon involves specifying both the set of possible outcomes and its intrinsic part (the structure of the experiment can be obtained from these).

The selection of relations for the systems in $\mathbf{K}_0$, is determined by what we consider relevant and identifiable in principle. Thus, in the case of these examples, two samples $S$ and $S'$ that coincide in the number of individuals that are in $R_0, R_1, \ldots, R_{n-1}$ but that differ with respect to the particular elements of $P$ that belong to each, are considered different. Therefore, an event might be the occurrence of a particular outcome $\mathfrak{A}_S$. In order to obtain this event, we have to include in the systems of $\mathbf{K}_0$ as distinguished elements, all members of $P$.

In summary, the simple probability structure describing the choosing of a sample of size $m$ from a population $P$ with relevant properties $R_0, \ldots, R_{n-1}$, is given by the pair $K_0 = \langle \mathbf{K}_0, \mathfrak{B}_0 \rangle$, where $\mathbf{K}_0 = \{\mathfrak{A}_S : S$ is a subset of $P$ of $m$ elements$\}$ and $\mathfrak{B}_0$, the intrinsic part, is $\langle P, R_0, \ldots, R_{m-1}, a \rangle_{a \in P}$.

Example 2. Distribution of $r$ balls into $n$ cells.

There are several ways of setting up the simple probability structures depending on the statistics to be obtained. Which of these applies, depends on the evidence available.

2.1) Maxwell-Boltzmann statistics. In this case, the balls are identifiable. We could construct a structure similar to that of Example 1. The relational systems would be of the form $\langle A, P_0, \ldots, P_{n-1}, a \rangle_{a \in A}$ with $P_i$ being the balls in cell $i$ in the particular outcome considered. The intrinsic part would be $\langle A, a \rangle_{a \in A}$ and the structural part $\langle A, P_0, \ldots, P_{n-1} \rangle$. However, in order to give a uniform account of the statistics involved, I will take a different course. The set $\mathbf{K}_1$ of possible outcomes consists of all systems $\mathfrak{A}_O = \langle A, P_0, \ldots, P_{n-1}, B_0, \ldots, B_{r-1}, O \rangle$ where $A = \{\langle i, m \rangle : i < r$ and $m < n\}$ (i.e., $A = r \times n$); $P_m$ consists of all pairs with $m$ as second coordinate (i.e., $P_m = \{\langle i, m \rangle : i < r\}$); $B_i$ consists of all pairs with $i$ as first coordinate (i.e., $B_i = \{\langle i, m \rangle : m < n\}$); and $O$ is any subset of $A$ that is a function with domain the numbers less than $r$.

Each element of $A$, $\langle i, m \rangle$ represents the possibility that ball $i$ might be in cell $m$. $P_0, \ldots, P_{n-1}$ stand for the $n$ cells, and $B_0, \ldots, B_{r-1}$ stand for the $r$ balls. $O$ is the particular partition in the outcome represented by $\mathfrak{A}_O$; it has to be a function, because each ball can be only in one cell.

It is clear that the intrinsic part of $K_1$ should be $\mathfrak{B}_1 = \langle A, P_0, \ldots, P_{n-1}, B_0, \ldots, B_{r-1} \rangle$ and $\mathfrak{A}_{O,\,st} = \langle A, O \rangle$. Thus, our simple probability structure is $K_1 = \langle \mathbf{K}_1, \mathfrak{B}_1 \rangle$.

2.2) Böse-Einstein statistics. In this case, the balls are not identifiable. Thus, the systems in $\mathbf{K}_2$ for this statistic should be of the form $\mathfrak{C}_O = \langle A, P_0, \ldots, P_{n-1}, O \rangle$ with $A, P_0, \ldots, P_{n-1}$ the same as for 2.1. $O$, however, has to be different. Since the balls are not identifiable, $B_0,\dots,B_{r-1}$ do not appear. In this statistics we do not really have balls, but numbers of positions in the different cells being occupied or not occupied. Thus, the pair $\langle i,m\rangle$ in $A$ represents the possibility, not of the ball $i$ being in the cell $m$, but of the position $i$ in the cell $m$ being occupied. Also, if position $i$ in cell $m$ is occupied (i.e., $\langle i,m\rangle \in O$), then all positions $j < i$, should also be occupied in cell $m$. That is, $O$ has $r$ elements and the property that $\langle i,m\rangle \in O$ and $j < i$, implies $\langle j,m\rangle \in O$. The intrinsic part is $\mathfrak{B}_2 = \langle A, P_0,\dots,P_{n-1}\rangle$. More formally, the experiment is described by $K_2 = \langle \mathbf{K}_2, \mathfrak{B}_2\rangle$, where $\mathbf{K}_2 = \{\langle A, P_0,\dots,P_{n-1}, O\rangle : O$ has $r$ elements, and for every $m < n$, there is an $i < r$ such that $O^{-1*}\{m\} = i\}$. Here, $O^{-1*}\{m\} = \{p : \langle p,m\rangle \in O\}$, and $i$ is identified with the set of numbers less than $i$.

2.3) Fermi-Dirac statistics. Balls are not identifiable and each cell has at most one ball. The systems in $\mathbf{K}_3$ are similar to those in $\mathbf{K}_2$ with the difference that the cells $P_0,\dots,P_{n-1}$ contain only one position, i.e., $A = \{\langle 0,m\rangle : m < n\}$, $P_m = \{\langle 0,m\rangle\}$, and $O$ is a subset of $A$ of $r$ elements.

$\mathfrak{B}_3$, the intrinsic part, is $\langle A, P_0,\dots,P_{n-1}\rangle$.

Example 3. A circular roulette over which a fixed force is applied, but starting from a variable position. The outcome might be any point in a circle and results from starting at a particular position. The systems in $\mathbf{K}_4$ (the set of possible outcomes), may be taken to be of the form:

$$\mathfrak{A}_O = \langle C, r, f, P_{a,b}, O\rangle_{\{a,b\} \in D}\,.$$

$C$ is the set of points of the circle, and $D = \{\{a,b\} : a,b \in C$ and $a$ is not opposite to $b$, i.e. $a$ and $b$ are not the end points of a diameter$\}$. For each fixed $a,b \in D$, $P_{a,b}$ is the property that applies to an $x$ in $C$, if $x$ is in the shorter arc determined by $a,b$, i.e. $P_{a,b} = \{x : x \in C$ and $x$ is in the shorter arc determined by $a,b\}$. $r$ is the ternary operation defined by: $r(a,b,c)$ is the point $c$ rotated through the angle determined by $a$ and $b$. $f$ is the continuous unary function that associates each initial position with a final position, i.e. if the roulette starts at $x$, then it stops at $f(x)$. $O$ is the set containing the initial position (i.e. a one-element subset of $C$, varying with each outcome).

The structural part of $\mathfrak{A}_O$, which gives the structure of the experiment is $\mathfrak{A}_{O,\mathrm{st}} = \langle C, r, O\rangle$. $O$ is variable for each $\mathfrak{A}_O$; however $r$ is the same. Thus, we cannot distinguish in this case, as in the previous ones, the structural part as consisting of the variable relations of the systems in $\mathbf{K}_4$.

Notice that a simple set $C$ is not enough for defining a circle. We also need some operations. (I have chosen $r$ for its simplicity, but others might do.)

Therefore, since the experiment essentially involves a circle we need to put enough structure in $\mathfrak{A}_{0,\mathrm{st}}$ so as to insure that we are really dealing with a circle. If the structural part were just $\langle C, O\rangle$, the experiment would consist of the choosing of a point from the set $C$ and not from the circle $\langle C, r\rangle$.

The intrinsic part is $\mathfrak{B}_4 = \langle C, f, P_{a,b}\rangle_{\{a,b\}\in D}$. Thus, it contains all relations not in the structural part. This is the main reason for putting a relation in the intrinsic part. However, the relations in $\mathfrak{B}_4$ are, in a way, intrinsic to the elements of $C$. Thus, it is an intrinsic property of a point $x \in C$ to be or not to be in a certain arc $P_{ab}$, or to yield a final position $f(x)$. Notice that $f$ is a property of the roulette, and with this $f$ we can express the asymmetry of the roulette. For symmetric roulettes, we do not need $f$.

The relations in the intrinsic part are used to determine the σ-field of events $\mathcal{B}$. As will be seen in Section 4, $\mathcal{B}$ is determined by all the relations in the systems in **K**, i.e. by everything we think relevant and identifiable.

We remark that in this particular example, we would get the same $\mathcal{B}$ (see Section 4), with a countable number of $P_{ab}$, namely, we could just have $P_{ab}$'s for all $a, b$ in a set $C'$ dense in $C$.

We see, in this case, that in order to decide which is the structural part of an experiment we cannot rely just on mathematical features of **K** (as, for instance, fixed and variable relations in **K**). We also have to decide which of the common relations of the systems in **K** are necessary to insure that the experiment really is what it pretends to be. This decision is based on physical reasons and not just mathematical ones.

From these examples, we see that in order to describe the simple probability structures we need to specify the set of possible outcomes and the structure of the experiment, i.e. which are the structural parts of each outcome. However, since it is easier, and equivalent, to specify the intrinsic part, of which there is just one for all of **K**, I shall follow this course. Thus, a simple probability structure is given by a set of possible outcomes and its intrinsic part.

We arrive, thus, at the following definition:

A *simple probability structure* is a pair $K = \langle \mathbf{K}, \mathfrak{B}\rangle$ such that **K** is a non empty set of relational systems of a fixed similarity type (called the *set of outcomes of* $K$) and $\mathfrak{B}$ is a relational system (called the *intrinsic part of* $K$) satisfying $\mathfrak{A} \upharpoonright I^{\mathfrak{B}} = \mathfrak{B}$, for every $\mathfrak{A} \in \mathbf{K}$.

Notice that this definition implies that all systems in **K** have a common universe.

## 4. DEFINITION OF THE $\sigma$-FIELD OF EVENTS

Let us begin by looking at the events determined by a single relational system $\mathcal{A}$, i.e. an individual outcome. In this case, there are just two possible events: the occurrence or the non-occurrence of the outcome. We should be able to obtain these two events from $\mathcal{A}$ by operating in a natural way on the relations that form it.

Suppose that $A$ is the universe of the real world.

The occurrence of the outcome $\mathcal{A}$, with universe $A$, can be considered as the proposition that what is valid in $\mathcal{A}$, is realized (i.e., true in the real world). That which is valid in $\mathcal{A}$ is that which is satisfied in $\mathcal{A}$ by all the elements of $A$. Contrariwise, the non-occurrence of $\mathcal{A}$ can be considered as the proposition that what is impossible in $\mathcal{A}$, is realized. That which is impossible in $\mathcal{A}$ is that which is satisfied in $\mathcal{A}$ by no element of $A$. (i.e. by the empty set). Thus, in order to obtain the events 'occurrence of $\mathcal{A}$' and 'non-occurrence of $\mathcal{A}$' we should aim at defining 'valid in $\mathcal{A}$' and 'impossible in $\mathcal{A}$' from the relations that form $\mathcal{A}$.

Suppose that $\mathcal{A}$ is a relational system of the form $\langle A, R_i^{\mathcal{A}}\rangle_{i\in I}$, with $A$, the real world. The relations that constitute $\mathcal{A}$ are the $R_i^{\mathcal{A}}$'s. An n-ary relation $R_i^{\mathcal{A}}$ can be thought of, as the set of $n$-tuples of $A$ that would satisfy $R_i$ if the real world were $\mathcal{A}$. Hence, each $R_i^{\mathcal{A}}$ can be taken as a satisfaction set (in $\mathcal{A}$). The $R_i^{\mathcal{A}}$'s are the basic satisfaction sets for $\mathcal{A}$ and they contain all the information about the world included in the assertion that $\mathcal{A}$ occurs. Therefore, we should try to get 'valid in $\mathcal{A}$' (which represents $\mathcal{A}$ occurs) and 'impossible in $\mathcal{A}$' (which represents $\mathcal{A}$ does not occur) from the $R_i^{\mathcal{A}}$'s through the use of purely logical operations. But we saw above that 'valid in $\mathcal{A}$' corresponds to the satisfaction set $A$ and 'impossible in $\mathcal{A}$', to $\emptyset$. Thus, we should be able to arrive through natural logical operations from the $R_i^{\mathcal{A}}$'s at these two extreme satisfaction sets. This will be accomplished by defining an algebra of satisfaction sets (in $\mathcal{A}$), the satisfaction algebra for $\mathcal{A}$, in which the two distinguished elements $A$ and $\emptyset$ represent 'valid in $\mathcal{A}$' and 'impossible in $\mathcal{A}$'.

Consider, now, a simple probability structure $K = \langle \mathbf{K}, \mathfrak{B}\rangle$, in which all relational systems in $\mathbf{K}$ have universe $A$. Events, here, are certain subsets of $\mathbf{K}$. Similarly as above, the occurrence of an event $\mathbf{A} \subseteq \mathbf{K}$ can be taken as the proposition that what is valid in $\mathbf{A}$ (i.e. valid in all relational systems $\mathcal{A} \in \mathbf{A}$) is realized. On the contrary, the non-occurrence of $\mathbf{A}$ can be considered as the proposition that what is impossible in $\mathbf{A}$ (i.e. impossible in all relational systems $\mathcal{A} \in \mathbf{A}$) is realized. Since $\mathbf{K}$ exhausts all possibilities, the non-occurrence of $\mathbf{A}$ is equivalent to the occurrence of $\mathbf{K} - \mathbf{A}$; thus, what is valid in $\mathbf{A}$ is impossible in $\mathbf{K} - \mathbf{A}$, and vice-versa.

It is clear that, if what is valid in an $\mathfrak{A} \in \mathbf{A}$ is realized, then what is valid in $\mathbf{A}$ is realized. Also, if what is valid in an $\mathfrak{A} \notin \mathbf{A}$ is realized, then what is impossible in $\mathbf{A}$ (because it is valid in $\mathbf{K}-\mathbf{A}$) is realized. Thus, $\mathbf{A}$ occurs if and only if there is an $\mathfrak{A} \in \mathbf{A}$ that occurs.

Let us look at an example. Suppose that $\mathbf{K}$ consists of the outcomes of the experiment of casting a symmetric six-faced die. The proposition that the real outcome is even, is valid in the possible outcomes 2, 4 and 6. Hence, if $\mathbf{A}$ consists of these otucomes, and the proposition that the real outcome is even is realized, then $\mathbf{A}$ occurs. Thus, for $\mathbf{A}$ to occur the real outcome should be 2, 4 or 6, and, hence, one of these three possible outcomes should occur.

Thus, we should aim at defining 'valid in $\mathbf{A}$', for the subsets $\mathbf{A}$ of $\mathbf{K}$ that we consider to be events. As above, this will be done by defining an algebra that represents 'satisfaction in $\mathbf{K}$', *the satisfaction algebra for* $\mathbf{K}$. This algebra is obtained from the satisfaction algebras for the relational systems $\mathfrak{A}$ in $\mathbf{K}$. The satisfaction algebra for $\mathbf{K}$ contains special elements representing 'valid in $\mathbf{A}$' (and 'impossible in $\mathbf{A}$') for certain subsets $\mathbf{A}$ of $\mathbf{K}$. Precisely these $\mathbf{A}$'s are to be our events.

We start by defining the satisfaction algebra for a single relational system $\mathfrak{A} = \langle A, R_i^{\mathfrak{A}} \rangle_{i \in I}$. We have that $R_i^{\mathfrak{A}} \subseteq {}^nA$, for a certain $n$; but for different $i \in I$ the corresponding $n$'s are different. However, in order to operate freely, we need a uniform unit set from which all our satisfaction sets should be subsets. Thus, we must have as unit set, at least, ${}^{\omega}A$, the set of infinite (denumerable) sequences of elements of $A$. So, we take ${}^{\omega}A$ as our unit set (i.e. representing the event of the occurrence of $\mathfrak{A}$), and replace the $R_i^{\mathfrak{A}}$'s by

$$\mathbf{R}_i^{\mathfrak{A}} = \{x : x \in {}^{\omega}A, \text{ and } x \restriction \nu(i) \in R^{\mathfrak{A}}\}.$$

Recall that $\nu(i)$ is the arity of $R_i^{\mathfrak{A}}$; hence $x \restriction \nu(i)$ is $\langle x_0, x_1, \ldots, x_{\nu(i)-1} \rangle$, i.e. $x$ restricted to $\nu(i)$.

In the satisfaction algebra for $\mathfrak{A}$ that will be defined, ${}^{\omega}A$ represents 'valid in $\mathfrak{A}$' (i.e. satisfied by all elements of $A$), and $\emptyset$ represents 'impossible $\mathfrak{A}$' (i.e. satisfied by no elements of $A$).

The operations to be performed on the $\mathbf{R}_i^{\mathfrak{A}}$ should have a purely logical character, independent of the particular $\mathfrak{A}$. Since I do not know of any mathematical characterization of logical operations that is wholly satisfactory, the choosing of these operations may seem somewhat arbitrary. However, there is a more or less general agreement that the following list of operations and subsets of ${}^{\omega}A$ have a purely logical character.

(a) Diagonal subsets of ${}^{\omega}A$ of the form

$$\mathbf{D}_{\kappa\lambda} = \{x : x \in {}^{\omega}A, \text{ and } x_{\kappa} = x_{\lambda}\}.$$

(If $x \in {}^{\omega}A$, we denote by $x_{\kappa}$ the $\kappa^{th}$ component of $x$.)

(b) Boolean operations of finite unions, intersections, and complement.

(c) Cylindrifications. For each $\kappa < \omega$ we have the operation $C_{\kappa}$ that consists on cylindrification by translation parallel to the $\kappa^{th}$ axis of the space. More formally, if $\mathfrak{s} \in {}^{\omega}A$ and $u \in A$, let $\mathfrak{s}_u^{\kappa}$ be the member of ${}^{\omega}A$ such that $(\mathfrak{s}_u^{\kappa})_{\lambda} = \mathfrak{s}_{\lambda}$ if $\lambda \neq \kappa$, while $(\mathfrak{s}_u^{\kappa})_{\kappa} = u$. Then, if $\mathbf{X} \subseteq {}^{\omega}A$

$$C_{\kappa}\mathbf{X} = \{\mathfrak{s} : \mathfrak{s} \in {}^{\omega}A \text{ and } \mathfrak{s}_u^{\kappa} \in \mathbf{X} \text{ for some } u\}.$$

A field of subsets of ${}^{\omega}A$ that contains the diagonal elements and is closed under cylindrifications $C_{\kappa}$, for each $\kappa \in \omega$, is called a cylindric field of sets. If $F$ is a cylindric field of sets, the algebra $\langle F, \cup, \cap, \sim, \emptyset, {}^{\omega}A, C_{\kappa}, \mathbf{D}_{\kappa\lambda}\rangle_{\kappa\, \lambda < \omega}$ is a Cylindric set algebra of dimension $\omega$ and base set $A$. Cylindric set algebras and related algebras are thoroughly studied in Henkin, Monk, and Tarski 1981, for short HMT 81.

Since we want a $\sigma$-field of sets, we need to add countable operations of union and intersection. There are two alternative ways of closing our field of events under these countable operations. The first is to require a limited form of $\sigma$-completeness in the cylindric algebra. The second, which will be explained later, is not to ask for $\sigma$- completeness in the cylindric algebra, but to first obtain a field of events and then close it to the least $\sigma$-field that contains it. The resulting $\sigma$-fields are not, in general, the same with these two constructions. I believe that the first way is much preferable and, hence, it will be adopted officially as the construction of the $\sigma$-field of events. The second way will only be presented as an alternative. For the first way we need to add the following operations.

(d) Define for $\mathbf{X} \subseteq {}^{\omega}A$, the dimension set of $\mathbf{X}$ to be

$$\Delta\,\mathbf{X} = \{\kappa : C_{\kappa}X \neq \mathbf{X}\}.$$

A set $\mathbf{X} \subseteq {}^{\omega}A$ is 0-dimensional if $\Delta\mathbf{X} = \emptyset$. It is clear that $\emptyset$ and ${}^{\omega}A$ are 0-dimensional. If an $\mathbf{X}$ were such that $\Delta\,\mathbf{X}$ were infinite, we could never arrive at a 0-dimensional set by cylindrifications, since we would need to apply to $\mathbf{X}$ infinitely many cylindrifications and this is not allowed among our operations. Since we are interested in arriving at the 0-dimensional sets $\emptyset$ or ${}^{\omega}A$, we have no use for infinite-dimensional sets. Hence, we limit ourselves to sets $\mathbf{X}$ with $\Delta\mathbf{X}$ finite. There is no problem in this, since our initial sets $\mathbf{R}_i^{\mathcal{A}}$ and $\mathbf{D}_{\kappa\lambda}$ are finite-dimensional. A cylindric algebra in which every element is finite-dimensional is called *locally finite-dimensional*. We now define the countable operations we want. Let $\langle \mathbf{X}_i : i \in \omega\rangle$ be a countable family of subsets of ${}^{\omega}A$ for which $\cup\{\Delta\mathbf{X}_i : i\in\omega\}$ is finite; then we have $\cup\{\mathbf{X}_i : i\in\omega\}$ and $\cap\{\mathbf{X}_i : i\in\omega\}$ in $F$. A field of subsets of ${}^{\omega}A$ that is closed under these operations will be called a *locally finite-dimensional cylindric σ-field of sets* and the corresponding algebra, *a locally finite-dimensional cylindric set σ-algebra.*

Thus, we have completed the description of the operations that are needed.

(I) First construction of the σ-field of events

The operations described in (a)-(d) are preserved under intersections of algebras. Hence, we consider the least locally finite-dimensional cylindric σ-field of subsets of ${}^{\omega}A$ generated by $\{\mathbf{R}_i^{\mathcal{A}} : i\in I^{\mathcal{A}}\}$ and call it $C_{\mathcal{A}}$. The corresponding locally finite-dimensional cylindric set σ-algebra will be denoted $\mathfrak{C}_{\mathcal{A}}$; i.e. $\mathfrak{C}_{\mathcal{A}} = \langle C_{\mathcal{A}}, \cup, \cap, \sim, \emptyset, {}^{\omega}A, C_{\kappa}, \mathbf{D}_{\kappa\lambda}\rangle_{\kappa,\lambda\in\omega}$. We shall call $C_{\mathcal{A}}$ and $\mathfrak{C}_{\mathcal{A}}$ the locally finite-dimensional σ-field of sets or cylindric σ-algebra corresponding to the relational system $\mathcal{A}$. Our task, now, is to prove that the only 0-dimensional elements of $C_{\mathcal{A}}$ are $\emptyset$, and ${}^{\omega}A$. For the sake of completeness, the proof shall be given in its entirety, although much of it can be deduced from HMT 81 pp. 53, 57-59. First a definition (cf. HMT 81 p.6). Let $\mathfrak{F} = \langle F, \cup, \cap, \sim, \emptyset, {}^{\omega}A, C_{\kappa}, D_{\kappa\lambda}\rangle_{\kappa,\lambda\in\omega}$ be a cylindric set algebra; an element $\mathbf{X}\in F$ is *regular* provided that $g\in\mathbf{X}$, whenever $f\in\mathbf{X}$, $g\in{}^{\omega}A$ and $f\restriction\Delta\mathbf{X} = g\restriction\Delta\mathbf{X}$. $F$ and $\mathfrak{F}$ are called *regular* if each $\mathbf{X}\in F$ is regular.

It is clear that $\mathbf{R}_i^{\mathcal{A}}$ and $\mathbf{D}_{\kappa\lambda}$ are regular in $C_{\mathcal{A}}$. It is also easy to prove the following:

Theorem (HMT 81, I.4.3). *If $\mathfrak{F}$ is a regular cylindric set algebra then the only 0-dimensional elements are $\emptyset$ and ${}^{\omega}A$.*

Thus, what we want will follow from:

Theorem. *If $\mathfrak{F}$ is the least locally finite-dimensional cylindric σ-algebra containing a family $D$ of regular elements of finite dimension $\mathbf{X}\subseteq{}^{\omega}A$, then $\mathfrak{F}$ is regular.* (cf. HMT 81, I.4.1).

Proof. We shall use the following lemma (HMT 81, p.53).

(*) Let $\Gamma$ be a finite subset of $\omega$, $\mathbf{X}\in F$. Suppose that for all $f, g$, if $f\in\mathbf{X}$, $g\in{}^{\omega}A$ and $f\restriction(\Delta\mathbf{X}\cup\Gamma) = g\restriction(\Delta\mathbf{X}\cup\Gamma)$, then $g\in\mathbf{X}$. Then $\mathbf{X}$ is regular.

In order to prove (*), assume its hypothesis, and suppose that $f \in X$, $g \in {}^{\omega}A$, and $f \upharpoonright \Delta X = g \upharpoonright \Delta X$; we have to show that $g \in X$. Let $\Theta = \Gamma \sim \Delta X$, $f' = f \upharpoonright (\omega \sim \Theta) \cup (\Theta \times \{f_0\})$, $g' = g \upharpoonright (\omega \sim \Theta) \cup (\Theta \times \{f_0\})$. It is clear that $f', g' \in {}^{\omega}A$: the set $\{\kappa : f_\kappa \neq f'_\kappa\}$ is a finite subset of $\omega \sim \Delta X$, call it $\{\kappa_0, \ldots, \kappa_{n-1}\}$; since $f \in X$, we have that $f' \in C_{\kappa_0} C_{\kappa_1} \ldots C_{\kappa_{n-1}} X = X$. Clearly $f' \upharpoonright (\Delta X \cup \Gamma) = g' \upharpoonright (\Delta X \cup \Gamma)$, so by the hypothesis of (*), $g' \in X$. Finally, $\{\kappa : g_\kappa \neq g'_\kappa\}$ is a finite subset of $\omega \sim \Delta X$, so, as above, we prove that $g \in X$, as desired.

The proof of the theorem is the following. Let $R$ be the set of all finite-dimensional regular elements of $\mathfrak{F}$; it suffices to show that $R$ is a locally finite-dimensional cylindric $\sigma$-field of sets. Clearly $D_{\kappa\lambda} \in R$ for all $\kappa, \lambda < \omega$, and clearly $R$ is closed under $\sim$, since $\Delta X = \Delta (\sim X)$.

Now let $X_i \in R$ for every $i \in \omega$ and suppose $\cup \{\Delta X_i : i \in \omega\}$ is finite. We shall verify (*) with $\cup \{\Delta X_i : i \in \omega\} = \Gamma$. It is clear that $\Delta(\cap \{X_i : i \in \omega\} \subseteq \Gamma$.

Suppose $f \upharpoonright \Gamma = g \upharpoonright \Gamma$, $f \in \cap \{X_i : i \in \omega\}$ and $g \in {}^{\omega}A$. Then $f \upharpoonright \Delta X_i = g \upharpoonright \Delta X_i$ for every $i \in \omega$. Since the $X_i$'s are regular and $f \in X_i$, $g \in X_i$ for every $i \in \omega$. Hence $g \in \cap \{X_i : i \in \omega\}$.

Finally, suppose $X \in R$ and $\kappa < \omega$; we show that $C_\kappa X \in R$.

To this end we verify (*) with $\Gamma = \Delta X \supseteq \Delta C_\kappa X$. So, suppose $f \upharpoonright \Delta X = g \upharpoonright \Delta X$, $f \in C_\kappa X$ and $g \in {}^{\omega}A$. Then for some $a \in A$ we have $f^\kappa_a \in X$. Thus, $f^\kappa_a \upharpoonright \Delta X = g^\kappa_a \upharpoonright \Delta X$ and $g^\kappa_a \in {}^{\omega}A$, hence by the regularity of $X$, $g^\kappa_a \in X$. Thus, $g \in C_\kappa X$, as desired.

We now proceed to the definition of the locally finite-dimensional cylindric $\sigma$-algebra $\mathfrak{C}_K$ corresponding to the simple probability structure $K = \langle K, \mathfrak{B} \rangle$. We assume that the elements of $K$ have the form $\mathcal{A} = \langle A, R_i^{\mathcal{A}} \rangle_{i \in I}$, with a common universe $A$. For this definition we only need the set of outcomes $K$. $\mathfrak{C}_K$ will be a subdirect product of the algebras $\mathfrak{C}_{\mathcal{A}}$, for $\mathcal{A} \in K$. That is, the universe of $\mathfrak{C}_K$ is a subset $C_K$ of the direct product $\Pi \langle C_{\mathcal{A}} : \mathcal{A} \in K \rangle$ such that each projection $\pi_{\mathcal{A}}$ (i.e. $\pi_{\mathcal{A}}(x) = x_{\mathcal{A}}$, for $x \in \Pi \langle C_{\mathcal{A}} \;\; \mathcal{A} \in K \rangle$) is a homomorphism of $\mathfrak{C}_K$ onto $\mathfrak{C}_{\mathcal{A}}$.

The definition of $\mathfrak{C}_K$ goes as follows:

Let $R_i$, for $i \in I$, be the system

$$R_i = \langle R_i^{\mathcal{A}} : \mathcal{A} \in K \rangle .$$ (see page 15 for the definition of $R_i^{\mathcal{A}}$).

For elements of the form $\langle X^{\mathcal{A}} : \mathcal{A} \in K \rangle$ with $X^{\mathcal{A}} \subseteq {}^{\omega}A$ for every $\mathcal{A}$, define the Boolean and cylindric operations coordinatewise. That is,

$$\langle X^{a} : a \in \mathbf{K} \rangle + \langle Y^{a} : a \in \mathbf{K} \rangle = \langle X^{a} \cup Y^{a} : a \in \mathbf{K} \rangle ,$$
$$\langle X^{a} : a \in \mathbf{K} \rangle \cdot \langle Y^{a} : a \in \mathbf{K} \rangle = \langle X^{a} \cap Y^{a} : a \in \mathbf{K} \rangle ,$$
$$- \langle X^{a} : a \in \mathbf{K} \rangle = \langle \sim X^{a} : a \in \mathbf{K} \rangle ,$$
$$\mathbf{c}_{\kappa} \langle X^{a} : a \in \mathbf{K} \rangle = \langle C_{\kappa} X^{a} : a \in \mathbf{K} \rangle , \text{ for } \kappa \in \omega ,$$
$$\wedge \{ \langle X^{a}_{i} : a \in \mathbf{K} \rangle : i \in \omega \} = \langle \cap \{ X^{a}_{i} : i \in \omega \} : a \in \mathbf{K} \rangle ,$$
$$\vee \langle X^{a}_{i} : a \in \mathbf{K} \rangle : i \in \omega \} = \langle \cup \{ X^{a}_{i} : i \in \omega \} : a \in \mathbf{K} \rangle ,$$

and, finally,

$$\mathbf{d}_{\kappa\lambda} = \langle \mathbf{D}_{\kappa\lambda} : a \in \mathbf{K} \rangle , \text{ for } \kappa, \lambda < \omega ,$$
$$1 = \langle {}^{\omega}A : a \in \mathbf{K} \rangle , \text{ and } 0 = \langle \emptyset : a \in \mathbf{K} \rangle .$$

With these operations and diagonal elements, the full direct product $\Pi \langle \mathfrak{C}_{a} : a \in \mathbf{K} \rangle$ is a cylindric algebra. However, we consider only the locally finite-dimensional cylindric σ-algebra generated by the elements $\mathbf{R}_{i}$ as $\mathfrak{C}_{\mathbf{K}}$ $(= \langle \mathcal{C}_{\mathbf{K}}, +, \cdot, -, 0, 1, \mathbf{c}_{\kappa}, \mathbf{d}_{\kappa\lambda} \rangle_{\kappa,\lambda<\omega})$.

The family consisting of the systems $\mathbf{R}_{i}$ contains all the information in $\mathbf{K}$. Thus, we should just close it under the operations discussed above to obtain all that we want. $\mathbf{R}_{i}$ is a sort of satisfaction system in $\mathbf{K}$, i.e. it is the system of all satisfaction sets $\mathbf{R}^{a}_{i}$. We should have in $\mathcal{C}_{\mathbf{K}}$ all satisfaction sets obtained from $\mathbf{R}_{i}$ and $\mathbf{d}_{\kappa\lambda}$ by the logical operations discussed above. Thus, $\mathfrak{C}_{\mathbf{K}}$ is the least locally finite-dimensional σ-cylindric algebra that contains the $\mathbf{R}_{i}$'s and $\mathbf{d}_{\kappa\lambda}$, and is closed under the operations +, ·, ∨,∧ , and $\mathbf{c}_{\kappa}$ for $\kappa, \lambda \in \omega$.

We know that the only 0-dimensional elements of $\mathfrak{C}_{a}$ are $\emptyset$ and ${}^{\omega}A$, and that these elements represent the events of $a$ not occuring, and of $a$ occuring respectively. Thus, since an element in the product is 0-dimensional if and only if every component is so, we have that the only 0-dimensional elements in $\mathfrak{C}_{\mathbf{K}}$ are systems consisting solely of $\emptyset$ and ${}^{\omega}A$; i.e. $\langle X^{a} : a \in \mathbf{K} \rangle$ is 0-dimensional if and only if $X^{a} = \emptyset$ or $X^{a} = {}^{\omega}A$, for every $a \in \mathbf{K}$. Such 0-dimensional elements represent the occurrence of all $a$ for which $X^{a} = {}^{\omega}A$ and the non-occurrence of those for which $X^{a} = \emptyset$.

An event $\mathbf{A}$ in $\mathbf{K}$ is a subset of $\mathbf{K}$. The occurrence of $\mathbf{A}$ (which is, simultaneously, the non-occurrence of $\mathbf{K} - \mathbf{A}$) can be represented, as we saw before, by that which is valid in $\mathbf{A}$ and impossible in $\mathbf{K} - \mathbf{A}$. That which is valid in $\mathbf{A}$ and impossible in $\mathbf{K} - \mathbf{A}$ is that which is valid in all relational systems $a \in \mathbf{K}$ that belong to $\mathbf{A}$ and impossible in those $a \in \mathbf{K}$ that do not belong to $\mathbf{A}$. Therefore, 'valid in $\mathbf{A}$ and impossible in $\mathbf{K} - \mathbf{A}$' is represented in our algebra $\mathfrak{C}_{\mathbf{K}}$ by the 0-dimensional system $\langle X^{a} : a \in \mathbf{K} \rangle$ such that $\{ a : X^{a} = {}^{\omega}A \} = \mathbf{A}$ (and, hence, $\{ a : X^{a} = \emptyset \} = \mathbf{K} - \mathbf{A}$).

Events should be the subsets of $\mathbf{K}$ obtained by our logical operations from the basic satisfaction systems $\mathbf{R}_i$. Thus, events naturally correspond to the 0-dimensional elements of $C_{\mathbf{K}}$.

Let us call $zdC_{\mathbf{K}}$ the set of 0-dimensional elements of $C_{\mathbf{K}}$. Then, our σ-field of events, which are subsets of $\mathbf{K}$, consists of the sets $\{\mathfrak{a} : \mathbf{X}^{\mathfrak{a}} = {}^{\omega}A\}$ for all systems $\langle \mathbf{X}^{\mathfrak{a}} : \mathfrak{a} \in \mathbf{K}\rangle$ in $zdC_{\mathbf{K}}$. That is,

Definition. $B_{\mathbf{K}} = \{\{\mathfrak{a}: \mathbf{X}^{\mathfrak{a}} = {}^{\omega}A\} : \langle \mathbf{X}^{\mathfrak{a}} : \mathfrak{a} \in \mathbf{K}\rangle \in zdC_{\mathbf{K}}\}$.

It is clear that $B_{\mathbf{K}}$ is isomorphic to $zdC_{\mathbf{K}}$, and, hence, is a σ-field of subsets of $\mathbf{K}$.

(II) Second construction

Let $\mathfrak{a} = \langle A, R_i^{\mathfrak{a}}\rangle$ be a relational system. Instead of $C_{\mathfrak{a}}$, the least locally finite-dimensional cylindric σ-field of subsets of ${}^{\omega}A$, we consider the least cylindric field of subsets of ${}^{\omega}A$ generated by $\{R_i^{\mathfrak{a}} : i \in I^{\mathfrak{a}}\}$ and call it $C'_{\mathfrak{a}}$. The corresponding cylindric set algebra, which is locally finite-dimensional, will be denoted $\mathfrak{C}'_{\mathfrak{a}}$. Thus, we close $\{R_i^{\mathfrak{a}} : i \in I^{\mathfrak{a}}\}$ under the operations (a)-(c) (and not (d)). It is clear that $C'_{\mathfrak{a}} \subseteq C_{\mathfrak{a}}$.

In the same way as for $C_{\mathfrak{a}}$ we can prove that the only 0-dimensional elements of $\mathfrak{C}'_{\mathfrak{a}}$ are $\emptyset$ and ${}^{\omega}A$.

Similarly as for the first construction, we define the cylindric algebra $\mathfrak{C}'_{\mathbf{K}}$ corresponding to the simple probability structure $K = \langle \mathbf{K}, \mathfrak{B}\rangle$, where the elements of $\mathbf{K}$ have the form $\mathfrak{a} = \langle A, R_i^{\mathfrak{a}}\rangle_{i \in I}$. $\mathfrak{C}'_{\mathbf{K}} = \langle C'_{\mathbf{K}}, +, \cdot, -, 0, 1, c_{\kappa}, d_{\kappa\lambda}\rangle_{\kappa,\lambda<\omega}$, is the cylindric algebra generated by the elements $\mathbf{R}_i$ (recall that $\mathbf{R}_i = \langle \mathbf{R}_i^{\mathfrak{a}} : \mathfrak{a} \in \mathbf{K}\rangle$. Again, in this case $C'_{\mathbf{K}}$ is closed under finite unions and intersections and not under the countable operations. It is clear that $\mathfrak{C}'_{\mathbf{K}}$ is a subdirect product of the algebras $\mathfrak{C}'_{\mathfrak{a}}$, for $\mathfrak{a} \in \mathbf{K}$ and that $C'_{\mathbf{K}} \subseteq C_{\mathbf{K}}$.

Now, if we consider

$\overline{B}_{\mathbf{K}} = \{\{\mathfrak{a}: \mathbf{X}^{\mathfrak{a}} = {}^{\omega}A\} : \langle \mathbf{X}^{\mathfrak{a}} : \mathfrak{a} \in \mathbf{K}\rangle \in zdC'_{\mathbf{K}}\}$ then $\overline{B}_{\mathbf{K}}$ is just a field of sets not closed under countable operations. Hence, we take $B'_{\mathbf{K}}$ as the least σ-field of subsets of $\mathbf{K}$ containing $\overline{B}_{\mathbf{K}}$. Since $\overline{B}_{\mathbf{K}}$ is contained in $B_{\mathbf{K}}$, we also have that $B'_{\mathbf{K}}$ is contained in $B_{\mathbf{K}}$.

We have, thus, obtained a σ-field of events $B_{\mathbf{K}}$ (or $B'_{\mathbf{K}}$) through the use of operations of a purely logical character acting upon the satisfaction systems $\mathbf{R}_i (= \langle \mathbf{R}_i^{\mathfrak{a}} : \mathfrak{a} \in \mathbf{K}\rangle)$, which contain all the information provided by $\mathbf{K}$.

I believe that $\mathcal{B}_{\mathbf{K}}$ is preferable to $\mathcal{B}'_{\mathbf{K}}$, because it seems more natural to obtain the σ-field of sets directly from the cylindric algebra $\mathfrak{C}_{\mathbf{K}}$, which is the natural structure obtained from $\mathbf{K}$. $\mathcal{B}'_{\mathbf{K}}$, on the other hand, is not obtained directly from $\mathfrak{C}'_{\mathbf{K}}$, but only after some closure operations. Another reason for this preference is discussed in Section 6.

Let us now consider the σ-field of events corresponding to the examples discussed in the previous section. In all cases where **K**, the set of outcomes, is finite (i.e. in Examples 1 and 2), $\mathcal{B}_{\mathbf{K}}$ (which is the same as $\mathcal{B}'_{\mathbf{K}}$) is the power set of **K**. This can be easily proved using a result of Section 6. In Example 3, $\mathcal{B}_{\mathbf{K}_4}$ and $\mathcal{B}'_{\mathbf{K}_4}$ consist of the sets $\{\mathfrak{A} : O^{\mathfrak{A}} \subseteq B\}$ for each $B$ that is a projective (for $\mathcal{B}_{\mathbf{K}_4}$) or a Borel (for $\mathcal{B}'_{\mathbf{K}_4}$) set generated by the arcs of the circle. Borel sets are well-known, i.e., in this case they constitute the least σ-field of subsets of the circle generated by the arcs. Projective sets, which include the Borel sets, not so well-known, so a brief description of them is in order. Analytic sets or A-sets are projections of Borel sets in two dimensions ($^2C$) over one dimension ($C$). CA-sets are complements of A-sets. Consider CA-sets in two dimensions (i.e. they are complements of projections of Borel sets in $^3C$ into $^2C$) and project them into $C$; thus, we obtain the PCA-sets. Continuing in this fashion we obtain the CPCA-sets, PCPCA-sets, etc. The union of all these classes constitute the class of projective sets. It can be shown that all Borel sets are A-sets but that not all A-sets are Borel, and that new sets appear in each of the classes CA, PCA, CPCA, etc. (see Kuratowski-Mostowski 1976 for the main properties of Borel and projective sets). A sketch of the proof that $\mathcal{B}_{\mathbf{K}_4}$ is the class of projective sets, and that, $\mathcal{B}'_{\mathbf{K}_4}$ is the class of Borel sets, appears in Section 6.

## 5. DEFINITION OF THE GROUP OF INVARIANCE

Let $K = \langle \mathbf{K}, \mathfrak{B} \rangle$ be a simple probability structure. Assume that the relational systems in **K** have a common universe $A$.

Let $\mathfrak{A} = \langle A, R_i \rangle_{i \in I}$ be a relational system and $g$ a permutation of $A$ (i.e. a one-one function of $A$ onto $A$). We define $g^*\mathfrak{A} = \langle A, g(R_i) \rangle_{i \in I}$, where $g(R_i) = \{\langle gx_0, \ldots, gx_{n-1} \rangle : \langle x_0, \ldots, x_{n-1} \rangle \in R_i\}$.

Let $\mathbf{B} \subseteq \mathbf{K}$; define $\mathbf{B}_{st} = \{\mathfrak{A}_{st} : \mathfrak{A} \in \mathbf{B}\}$. Recall that $\mathfrak{A}_{st} = \mathfrak{A} \restriction (I^{\mathfrak{A}} \sim I^{\mathfrak{B}})$, i.e. $\mathfrak{A}$ without the relations in $\mathfrak{B}$. Also, if $g$ is a permutation of $A$, define

$$\mathbf{B}^{g} = \{\mathfrak{A} : \text{there is an } \mathfrak{A}' \in \mathbf{B} \text{ such that } \mathfrak{A}_{st} = g^*(\mathfrak{A}'_{st})\}.$$

We want a group $G_K$ of permutations of $A$ under which the laws of the phenomenon, represented, in a way, by $K$, are invariant. $G_K$ should be a subgroup of the group of permutations of $A$, since the only distinctions among the elements of $A$ should come from $K$ itself.

Definition. $G_K$ is the group of all permutations $\sigma$ of the universe of $\mathbf{K}$, $A$, such that, for every $\mathbf{B} \in B_{\mathbf{K}}$, $\mathbf{B}^{\sigma}$ and $\mathbf{B}^{\sigma^{-1}} \in B_{\mathbf{K}}$.

This implies, in particular, that $\mathbf{K}^{\sigma}$ and $\mathbf{K}^{\sigma^{-1}} \subseteq \mathbf{K}$, i.e., $\sigma^*(\mathfrak{a}_{st}) = \mathfrak{a}'_{st}$ for some $\mathfrak{a}' \in \mathbf{K}$, if $\mathfrak{a} \in \mathbf{K}$.

The measure $\mu$ on $B_{\mathbf{K}}$, should be a measure such that $\mu(\mathbf{B}) = \mu(\mathbf{B}^{\sigma})$ for every $\mathbf{B} \in B_{\mathbf{K}}$ and $\sigma \in G_K$. This requirement is similar to Carnap's Axiom of Symmetry (see Carnap 1971, Ch. 9).

Passing, now, to the examples, we see that in Example 1, $G_{\mathbf{K}_0}$ consists of all permutations of the population of balls $P$.

The other examples, however, impose some restrictions on $G_{K_1}$, $G_{K_2}$, and $G_{K_3}$. Namely for $\sigma$ to be in $G_{K_1}$, $\sigma^*(O)$ must have the right character, according to which of $K_1$, $K_2$, $K_3$ is involved; for instance, in the case of $\mathbf{K}_1$, $\sigma^*(O)$ should be a function.

$K_4$, for Example 3, imposes another restriction on the permutations of $C$ that are in $G_{K_4}$. If $\sigma$ is a permutation of $C$, for $\sigma$ to belong to $G_{K_4}$, it is necessary that $\sigma^*(\langle C,r,O\rangle)$ be the structural part $\langle C,r,O'\rangle$ of another outcome. That is, $\sigma^*(\langle C,r\rangle) = \langle C,r\rangle$ ; hence $\sigma$ should be a rotation of the circle. This automatically insures that $\mathbf{B}^{\sigma} \in B_{\mathbf{K}_4}$ for $\mathbf{B} \in B_{\mathbf{K}_4}$, since $B_{\mathbf{K}_4}$ consists essentially of the projective sets (see Section 6 for a sketch of the proof of this fact).

The invariant measure exists and is unique in all of the examples. Examples 1 and 2 have a finite Boolean algebra in which all the atoms are equivalent under the corresponding group (i.e. for each pair of atoms $\mathbf{B}$, $\mathbf{B}'$, we have that $\mathbf{B}' = \mathbf{B}^{\sigma}$, for some $\sigma$ in the group). Hence, the invariant measure in essentially the counting measure.

The situation is more complicated for Example 3, i.e. the case of the roulette. Recall that $\mathfrak{B}_4$, i.e. the intrinsic part of this experiment, is $\langle C, f, P_{a,b}\rangle_{\{a,b\} \in D}$, where $f$ is a continuous unary function that associates each initial position with a final position. Recall that $\mathbf{A} \in B_{\mathbf{K}_4}$ (or $B'_{\mathbf{K}_4}$) if and only if $\mathbf{A} = \{\mathfrak{a} : O^{\mathfrak{a}} \in A\}$ for $A$ a projective set (or Borel set) of the circle. (Notice, for later use, that since $f$ is continuous, $A$ is projective (or Borel) if and only if $f^*(A)$ is.) The invariant measure $\mu_4$ is given by $\mu_4(\mathbf{A}) = \lambda(A)$, where $\mathbf{A}$ is as above and $\lambda$ is Lebesgue measure. Thus $\mathbf{A} = \{\alpha : 0^{\alpha} \in A\} \in \mathbf{B}_{\mathbf{K}_4}$ is $\mu_4$-measurable if and only if A is Lebesgue measurable. However, since it is implied by the Axiom of Constructibility (which is consistent with the usual axioms of set theory, Gödel 1939) that there are projective sets of real numbers which are not Lebesgue measurable, it cannot be proved from the usual axioms that all sets in $B_{\mathbf{K}_4}$ are $\mu_4$-measurable.

On the other hand, it is proved in Solovay 1970 that it is consistent with the usual axioms of set theory, including the axiom of choice, to assume that all projective sets are Lebesgue measurable. Also, adding certain strong axioms to the usual ones, such as the Axiom of Projective Determinacy (in favor of which there is some independent evidence, see Moschovakis 1980, Ch.6), it can be proved that all projective sets are Lebesgue measurable. It seems to me that the fact that the construction of $B_{\mathbf{K}_4}$ presented here is very natural as the construction for the σ-field of events, and that we should be able to assign probabilities to all events, give support to the acceptance of measurability for all projective sets, and indirectly, to the acceptance of these stronger axioms.

Since all Borel sets are provably Lebesgue measurable with the usual set-theoretical axioms, the same is true for sets in $B'_{\mathbf{K}_4}$ . This fact may be considered a reason in favor of the Second Construction of the σ-field of events. However, I believe that the arguments for $B_{\mathbf{K}_4}$ as the σ-field of events, outweigh this consideration.

Finally, a word should be said about the role played by the intrinsic part and the structural part of a simple probability structure **K**. Although both may have influence on $B_K$ and $G_K$, I would say that the main importance of the intrinsic part is in the determination of $B_{\mathbf{K}}$. The structural part, in its turn, has a major influence on $G_K$. In fact, in all our examples (1,2 and 3), $G_K$ is exclusively determined by the structure of the experiment. However, in general, there may be influence of the intrinsic part, since, according to the definition of $G_K$, $B_{\mathbf{K}}$ conditions this group. In spite of this possible influence of $B_{\mathbf{K}}$, I do not have an example in which the intrinsic part of $K$ plays a role in the definition of $G_K$.

## 6. PROBABILITY AND TRUTH

In order to model truth, and probability as a sort of "partial truth", we need to introduce a language as a mathematical object. (For a discussion of probability as "partial truth" see Chuaqui 1977a, 1979 and 1980.) Thus, we would have the language and its possible interpretations, the relational systems or simple probability structures, inside a mathematical theory, namely, set theory. Truth and probability appear as relations between the language and its interpretations. Thus we say 'true under such and such interpretation of the symbols of the language'; and the same, for probability.

The first complete mathematical definition of truth for formalized languages was given in Tarski 1935, and has had a decisive importance in the development of the Foundations of Mathematics.

The languages appropriate for dealing with $\sigma$-additive probability measures are called $\mathcal{L}_{\omega_1\omega}$-languages and admit conjunctions and disjunctions of countable sets of formulas. For a development of $\mathcal{L}_{\omega_1\omega}$-languages, see Keisler 1971.

Let us now proceed with the definition of the language (or better languages) $\mathcal{L}_{\omega_1\omega}$. This definition is a modification of that appearing in Keisler 1971, suitable for our purposes. Let $\langle I,\nu\rangle = \mathcal{s}$ be a similarity type, i.e. $I$ is an arbitrary set of indices and $\nu : I \to \omega$. We have an $\mathcal{L}_{\omega_1\omega}$-language $\mathcal{L}_{\mathcal{s}}$, for each similarity type $\mathcal{s}$. For each $i \in I$, we introduce a predicate symbol $P_i$ (In a complete formalization of the languages in set theory, symbols are represented by sets.) This symbol $P_i$ will denote the relation $R_i^{\mathfrak{A}}$ in the structure $\mathfrak{A}$. We also have as logical symbols a binary relation $=$ of equality; $\neg$, negation; $\exists$, $\forall$, the existential and universal quantifiers; $\bigwedge$, $\bigvee$, finite or countable conjunction and disjunction; and parentheses. Finally, we have a sequence $\langle \nu_\kappa : \kappa < \omega\rangle$ of variables.

The definition of a well-formed formula is the usual one (with one modification):

(i) An atomic formula is a predicate symbol $P_i$ followed by $\nu(i)$ variables, i.e. $P_i\, \nu_{k_0} \ldots \nu_{k\,\nu(i)-1}$; or an expression of the form $\nu_\kappa = \nu_\lambda$.

(ii) If $\phi$ and $\psi$ are formulas, and $\nu_\kappa$ a variable, then $(\neg\phi)$, $\exists\, \nu_\kappa \phi$, $\forall \nu_\kappa \phi$ are formulas.

A variable $\nu_\kappa$ occurs free at the place $k$ of a formula, if it is not in the scope of a quantifier $\exists\, \nu_\kappa$ or $\forall\, \nu_\kappa$ at that place. It occurs free in a formula, if it occurs free at some place. $\mathrm{Fv}\,\phi$, denotes the set of free variables of $\phi$.

(iii) If $\Phi$ is a finite or countable, non-empty, set of formulas such that $\cup\{\mathrm{Fv}\,\phi : \phi \in \Phi\}$ is finite, then $\bigwedge \Phi$ and $\bigvee \Phi$ are formulas.

There are two main differences with the languages in Keisler 1971. In the first place, we have a denumerable sequence of variables, instead of an $\omega_1$-sequence. In the second place, we restrict denumerable disjunctions and conjunctions only to cases where the set of formulas involved has a finite number of free variables. It is easy to modify the derivation system of Keisler 1971, so that a sentence $\phi$ is a consequence of a set $\Sigma$ of sentences in Keisler's system if and only if, it is a consequence in our system.

We also consider $\mathcal{L}_{\omega\omega}$-languages where only finite conjunctions and disjunctions are allowed (in clause (iii)).

For each simple probability structure, we need a particular $\mathcal{L}_{\omega_1\omega}$-language (or $\mathcal{L}_{\omega\omega}$-language). For instance in Example 1, we need a language with predicate symbols to be interpreted by $R_0,\ldots,R_{m-1}$ and $S$, let us say $P_0,\ldots,P_{m-1}$, and $P_m$.

The definition of truth and satisfaction for $\mathcal{L}_s$ an $\mathcal{L}_{\omega_1\omega}$-language or $\mathcal{L}'_s$ an $\mathcal{L}_{\omega\omega}$-language will be done in terms of satisfaction classes. First satisfaction and truth in a relational system $\mathfrak{A} = \langle A, R_i^{\mathfrak{A}} \rangle_{i\in I^{\mathfrak{A}}}$ of the same similarity type as $\mathcal{L}_s$. For each formula $\phi$ we define a subset of ${}^{\omega}A$, $\phi^{\mathfrak{A}}$ called the satisfaction class of $\phi$. $\phi^{\mathfrak{A}}$ contains the $\omega$-tuples that satisfy $\phi$ in $\mathfrak{A}$. The definition is by recursion on formulas.

(i) If $\phi$ is atomic of the form $P_i\, v_0 \cdots v_{\nu(i)-1}$, then $\phi^{\mathfrak{A}} = \mathbf{R}_i^{\mathfrak{A}}$. If $\phi$ is $v_\kappa = v_\lambda$, then $\phi^{\mathfrak{A}} = \mathbf{D}_{\kappa\lambda}$.

(ii) If $\phi$ is $(\neg\psi)$, then $\phi^{\mathfrak{A}} = \sim \psi^{\mathfrak{A}}$

(iii) If $\phi$ is $\exists v_\kappa \psi$, then $\phi^{\mathfrak{A}} = C_\kappa \psi^{\mathfrak{A}}$; if $\phi$ is $\forall v_\kappa \psi$, then $\phi^{\mathfrak{A}} = \sim C_\kappa \sim \psi^{\mathfrak{A}}$

(iv) If $\phi$ is $\bigwedge \Phi$, then $\phi^{\mathfrak{A}} = \cap \{\psi^{\mathfrak{A}} : \psi \in \Phi\}$.

If $\phi$ is $\bigvee \Phi$, then $\phi^{\mathfrak{A}} = \cup \{\psi^{\mathfrak{A}} : \psi \in \Phi\}$.

It is clear from the definition (see HMT 1981) that if $\mathcal{L}_s$ is an $\mathcal{L}_{\omega_1\omega}$-language then $\{\phi^{\mathfrak{A}} : \phi \text{ a formula in } \mathcal{L}_s\} = \mathcal{C}_{\mathfrak{A}}$, the locally finite-dimensional cylindric $\sigma$-field of sets corresponding to $\mathfrak{A}$. If $\mathcal{L}'_s$ is an $\mathcal{L}_{\omega\omega}$-language, then $\{\phi^{\mathfrak{A}} : \phi \text{ a formula in } \mathcal{L}'_s\} = \mathcal{C}'_{\mathfrak{A}}$, the cylindric field of sets corresponding to $\mathfrak{A}$.

We define $\mathfrak{A} \models \phi$ ($\phi$ *is valid in* $\mathfrak{A}$) for any formula $\phi$ in $\mathcal{L}_s$, by $\phi^{\mathfrak{A}} = {}^{\omega}A$. In case $\phi$ is a sentence (i.e. a formula without free variables), $\phi^{\mathfrak{A}} = {}^{\omega}A$ or $\phi^{\mathfrak{A}} = \emptyset$, i.e. $\phi$ is valid or impossible in $\mathfrak{A}$. For sentences, $\mathfrak{A} \models \phi$ is read '$\phi$ *is true in* $\mathfrak{A}$' or '$\mathfrak{A}$ *is a model of* $\phi$'.

The satisfaction class, $\phi^{\mathbf{K}}$, of a formula $\phi$ in the simple probability structure $K = \langle \mathbf{K}, \mathcal{L} \rangle$ of the same similarity type as the language $\mathcal{L}_s$, is defined by,

$$\phi^{\mathbf{K}} = \langle \phi^{\mathfrak{A}} : \mathfrak{A} \in \mathbf{K} \rangle .$$

We also define truth in $\mathbf{K}$ for sentences $\phi$, by

$$\mathbf{K} \models \phi, \text{ if and only if, } \phi^{\mathbf{K}} = \langle {}^{\omega}A : \mathfrak{A} \in \mathbf{K} \rangle .$$

It is also easy to see that for $\mathcal{L}_{\delta}$, an $\mathcal{L}_{\omega_1\omega}$-language, $\{\phi^{\mathbf{K}} : \phi$ a formula of $\mathcal{L}_{\delta}\} = \mathcal{C}_{\mathbf{K}}$, the domain of the locally finite-dimensional cylindric $\sigma$-algebra corresponding to $\mathbf{K}$. For $\mathcal{L}'_{\delta}$, and $\mathcal{L}_{\omega\omega}$-language, we have $\{\phi^{\mathbf{K}} : \phi$ a formula of $\mathcal{L}'_{\delta}\} = \mathcal{C}'_{\mathbf{K}}$. It is clear, that the 0-dimensional systems of $\mathcal{C}_{\mathbf{K}}$ or $\mathcal{C}'_{\mathbf{K}}$ are of the form $\phi^{\mathbf{K}}$, for $\phi$ a sentence of $\mathcal{L}_{\delta}$ or $\mathcal{L}'_{\delta}$, respectively. We also have, for sentences $\phi$, that

$$Mod_{\mathbf{K}}(\phi) = \{\mathfrak{A} : \mathfrak{A} \in \mathbf{K}, \phi^{\mathfrak{A}} = {}^{\omega}A\};$$

where $\phi^{\mathfrak{A}}$ is the $\mathfrak{A}$ th. component of the system $\phi^{\mathbf{K}} = \langle \phi^{\mathfrak{A}} : \mathfrak{A} \in \mathbf{K} \rangle$.

Also, $Mod_{\mathbf{K}}(\phi)$ is defined by,

$$Mod_{\mathbf{K}}(\phi) = \{\mathfrak{A} : \mathfrak{A} \in \mathbf{K}, \mathfrak{A} \models \phi\} .$$

In order to clarify the matter further by looking at Example 1, we can say that while the predicate $P_m$ (corresponding to $S$) is interpreted in $\mathfrak{A}$ by $\mathbf{S}^{\mathfrak{A}}$, it is interpreted in $\mathbf{K}$ by $\mathbf{S} = \langle \mathbf{S}^{\mathfrak{A}} : \mathfrak{A} \in \mathbf{K} \rangle$. This last system cannot be identified with any $\mathbf{S}^{\mathfrak{A}}$, since $\mathbf{S}^{\mathfrak{A}} \neq \mathbf{S}^{\mathfrak{A}'}$ for $\mathfrak{A} \neq \mathfrak{A}'$. On the other hand, the predicate $P_0$ (corresponding to $R_0$) is interpreted by $\mathbf{R}_0^{\mathfrak{A}}$ in $\mathfrak{A}$, and by $\mathbf{R}_0 = \langle \mathbf{R}_0^{\mathfrak{A}} : \mathfrak{A} \in \mathbf{K} \rangle$ in $\mathbf{K}$; but $\mathbf{R}_0^{\mathfrak{A}} = \mathbf{R}_0^{\mathfrak{A}'}$ for $\mathfrak{A} \neq \mathfrak{A}'$, and, hence, these two interpretations may be identified.

From the definition of $\mathcal{B}_{\mathbf{K}}$, we conclude by using the characterizations of $Mod_{\mathbf{K}}(\phi)$, that the $\sigma$-field $\mathcal{B}_{\mathbf{K}} = \{Mod_{\mathbf{K}}(\phi) : \phi$ a sentence of $\mathcal{L}_{\delta}\}$ (if $\mathcal{L}_{\delta}$ is an $\mathcal{L}_{\omega_1\omega}$-language). However, $\mathcal{B}'_{\mathbf{K}}$ is not, in general, $\{Mod_{\mathbf{K}}(\phi) : \phi$ a sentence of $\mathcal{L}'_{\delta}\}$, for $\mathcal{L}'_{\delta}$ an $\mathcal{L}_{\omega\omega}$-language. This last class is only a field of subsets of $\mathbf{K}$ and not, in general, a $\sigma$-field. $\{Mod_{\mathbf{K}}(\phi) : \phi$ a sentence of $\mathcal{L}'_{\delta}\} = \overline{\mathcal{B}}_{\mathbf{K}}$, where $\overline{\mathcal{B}}_{\mathbf{K}}$ was introduced in Section 4. $\mathcal{B}'_{\mathbf{K}}$ is, in its turn, the least $\sigma$-field of subsets of $\mathbf{K}$ containing $\overline{\mathcal{B}}_{\mathbf{K}}$.

This difference between $\mathcal{B}'_{\mathbf{K}}$ and $\mathcal{B}_{\mathbf{K}}$ is a strong argument in favor of the latter, since it can be characterized as the collection of sets of models of sentences of a language, while $\mathcal{B}'_{\mathbf{K}}$ cannot be so described, at least in a natural way.

In Examples 1 and 2, where the relational systems are finite, we have that for each $\mathfrak{A} \in \mathbf{K}_i$ (where $i$ = 0,1,2 or 3) there is a sentence $\phi$ of the appropriate language ($\mathcal{L}_{\omega_1\omega}$-languages coincide with $\mathcal{L}_{\omega\omega}$-languages, for these examples), such that for every $\mathfrak{A}' \in \mathbf{K}_i$,

$$\mathfrak{A}' \models \phi \text{ if and only if } \mathfrak{A}' = \mathfrak{A}.$$

Thus, the set $\{\mathfrak{A}\}$ is $Mod_{\mathbf{K}_i}(\phi)$ for a certain sentence $\phi$. This proves that $B_{\mathbf{K}_i}$ consists of all subsets of $\mathbf{K}_i$ for $i = 0,1,2,3$.

The analysis of Example 3 is more complicated. $B_{\mathbf{K}_4}$ is different from $B'_{\mathbf{K}_4}$, is this case. The difference stems from the fact more sets are definable in an $\mathcal{L}_{\omega_1\omega}$-language than in an $\mathcal{L}_{\omega\omega}$-language. Let us call the similarity type of $\mathbf{K}_4$, $s_4$ and the corresponding $\mathcal{L}_{\omega_1\omega}$-language and $\mathcal{L}_{\omega\omega}$-language, $\mathcal{L}_{s_4}$ and $\mathcal{L}'_{s_4}$. Since the relation $O$ contains exactly one element in each system in $\mathbf{K}_4$, it can be symbolized in the languages by an individual constant $a$ instead of a predicate. Thus, if $a^{\mathfrak{A}}$ is the denotation of $a$ in the system $\mathfrak{A}$, then $O^{\mathfrak{A}} = \{a^{\mathfrak{A}}\}$.

It is clear that a sentence $\phi(a)$ is true in a system $\mathfrak{A}$ in $\mathbf{K}_4$ if and only if $a^{\mathfrak{A}}$ (i.e. the denotation of $a$ in $\mathfrak{A}$) is contained in the subset $B$ of the circle defined by $\phi(x)$ where $a$ does not occur in $\phi(x)$ (i.e. $B$ is the set of $x \in C$ that satisfy $\phi(x)$, in symbols, $B = \phi(x)^{\mathfrak{A}}$). Thus, in order to determine which sets are in $B_{\mathbf{K}_4}$ and $B'_{\mathbf{K}_4}$ we just need to ascertain which subsets of the circle are definable in systems $\mathfrak{A}$ in $\mathbf{K}_4$ by formulas $\phi(x)$ with one free variable, and not containing $a$, of $\mathcal{L}_{s_4}$ or $\mathcal{L}'_{s_4}$, respectively.

Let us first see the case of $\mathcal{L}_{s_4}$. In this $\mathcal{L}_{\omega_1\omega}$-language, all Borel sets in ${}^nC$ are definable by quantifier free formulas. Thus, all projective subsets of $C$ are definable by formulas with one free variable.

Moreover, these are the only definable subsets of the circle by formulas with one free variable, since the open formulas define exactly the Borel subsets of ${}^nC$, for every $n$. This is so, because $r$ and $s$ are continuous. Thus, we prove that $B_{\mathbf{K}_4}$ is the class of all subsets of $\mathbf{K}_4$ of the form $\{\mathfrak{A} : O^{\mathfrak{A}} \subseteq B\}$, where $B$ is any projective subset of the circle.

The formulas with one free variable of $\mathcal{L}'_{s_4}$, on the other hand, are only able to define finite unions of arcs, because $s$ and $r$ send arcs onto arcs. Hence, $\bar{B}_{\mathbf{K}_4}$ consists of the subsets of $\mathbf{K}_4$ of the form $\{\mathfrak{A}: O^{\mathfrak{A}} \subseteq B\}$, where $B$ is any finite union of arcs of the circle. Since $B'_{\mathbf{K}_4}$ is the least $\sigma$-field containing $\bar{B}_{\mathbf{K}_4}$, $B'_{\mathbf{K}_4}$ consists of the subsets of $\mathbf{K}_4$ of the form $\{\mathfrak{A}: O^{\mathfrak{A}} \subseteq B\}$ where $B$ is a Borel set.

In order to facilitate the understanting of the assignment of the measure $\mu_4$, I shall give an example. Suppose that the formula $\phi(x)$, where $O$ does not appear, defines the subset of $C$, $B$ (i.e. $B = \{x : \langle C, r, f, P_{a,b}\rangle_{a,b \in D} \models \phi[x]\}$. In order to say that the result of the experiment is in $B$, we have to use the sentence $\psi$ given by

$$\exists x(O(x) \wedge \phi(f(x))) .$$

Then, the set $A \in B_{K_4}$ defined by $A = \mathrm{Mod}_{K_4}(\psi)$ is this event (i.e. the result is in $B$). We have, that $A = \{\mathfrak{A} : O^{\mathfrak{A}} \subseteq f^{-1*}B\}$. Thus, $\mu_4(A) = \lambda(f^{-1*}B)$, where $\lambda$ is Lebesgue measure.

We end this section with a definition of probability for sentences of $\mathcal{L}_s$ (or $\mathcal{L}'_s$) where $\mathcal{L}_s$ and $\mathcal{L}'_s$ are arbitrary $\mathcal{L}_{\omega_1\omega}$ or $\mathcal{L}_{\omega\omega}$-languages (not necessarily those appropriate for $K_4$). Let $K = \langle K, \mathcal{F}\rangle$ be a simple probability structure of similarity type $s$, and suppose that there is a unique measure $\mu$ on $B_K$ (or $B'_K$) invariant under $G_K$. Then for each sentence $\phi$ in $\mathcal{L}_s$ (or $\mathcal{L}'_s$), we define

$$P_K(\phi) = \mu(\mathit{Mod}_K(\phi)) .$$

Thus, $P_K(\phi)$ is a measure of the size of the set of models of $\phi$ that are in $K$. $P_K(\phi) = r$ may be interpreted as '$\phi$ is true in $K$ to the degree $r$'.

## 7. SOME MATHEMATICAL PROBLEMS ARISING FROM MY APPROACH TO PROBABILITY

Since what we need are measures invariant under certain groups, the main mathematical problems have to do with this. In particular, I shall mention the following problems.

### 7.1 Conditions of an algebraic character for the existence and uniqueness of the measure

I have already obtained necessary and sufficient conditions on a group and a $\sigma$-fields of sets for the existence of an invariant measure (Chuaqui 1977 b and 198+). In order to state this problem and see its relation to the simple probability structure, I will restate $B_K$ and $G_K$. Assume that $K = \langle K, \mathfrak{B}\rangle$ is a simple probability structure. Instead of $B_K$ consider $B_{st} = \{B_{st} : B \in B_K\}$. $B_{st}$ is a $\sigma$-field of subsets of $K_{st}$ isomorphic to $B_K$. Now, instead of $G_K$ consider $G^*_K = \{f^* : f \in G_K\}$, where $f^*$ is as defined in Section 5. It is clear that $G^*_K$ is a group of transformations of $B_{st}$. So if we obtain a measure on $B_{st}$ invariant under $G^*_K$ we can easily transfer it to a measure on $B_K$ invariant under the action of $G_K$.

Thus, the mathematical problem has been summarized as the obtaining of measures on a field of sets $F$ invariant under a group $G$.

Necessary and sufficient conditions for the existence of a finitely additive measure invariant under a group of transformations are given, for instance, in Tarski 1949. Here I shall concentrate on $\sigma$-additive probability measures.

In order to state the solutions found in my earlier papers, I need some definitions:

(1) $\mathbf{A} \simeq_G \mathbf{B}$ if and only if there are sequences $\mathbf{Y}_i$, $i \in \omega$ and $\mathbf{Z}_i$, $i \in \omega$ of elements of $F$ and $f_i$, $i \in \omega$ of elements of $G$ such that $\mathbf{A} = \Sigma_{i<\omega}\mathbf{Y}_i$, $\mathbf{B} = \Sigma_{i<\omega}\mathbf{Z}_i$, and $\mathbf{Z}_i = \mathbf{Y}_i^{f_i}$ for every $i \in \omega$.

Here $\Sigma_{i<\omega}\mathbf{Y}_i$ means the disjoint union of the $\mathbf{Y}_i$'s, i.e. $\Sigma_{i<\omega}\mathbf{Y}_i = \cup_{i<\omega}\mathbf{Y}_i$ if $\mathbf{Y}_i \cap \mathbf{Y}_j = \emptyset$ for $i \neq j$.

$\mathbf{A} \simeq_G \mathbf{B}$ can be read as: equivalent by countable decomposition under $G$.

(2) An element $\mathbf{A} \in F$ is $G$-negligible if there is a sequence $\mathbf{Y}_i$, $i \in \omega$ of pairwise disjoint elements of $F$ such that $\mathbf{A} \simeq_G \mathbf{Y}_i$, for every $i < \omega$.

It is clear that a measure $\mu$ invariant under $G$, has to be invariant under $\simeq_G$, and that a $G$-negligible set must have measure zero. Thus, $\mu$ has also to be invariant under the relation $\approx_G$, defined below.

(3) $\mathbf{A} \approx_G \mathbf{B}$ if there are $G$-negligible elements of $F$, $\mathbf{C}$, $\mathbf{D}$, $\mathbf{C}'$, $\mathbf{D}'$ and elements $\mathbf{A}'$, $\mathbf{B}'$ of $F$ such that $\mathbf{A}' \simeq_G \mathbf{B}'$, $\mathbf{A} \cup \mathbf{C} = \mathbf{A}' \cup \mathbf{C}'$, and $\mathbf{B} \cup \mathbf{D} = \mathbf{B}' \cup \mathbf{D}'$.

In case $F$ is a finite field of sets, there is always a measure invariant under a group. This measure is unique if and only if all atoms in $F$ are $\simeq_G$-equivalent (see Chuaqui 1977a, Section 4).

Now, some more definitions.

(4) Let $J \subseteq F$. We say that $\mu$ is $J$-positive if $\mu(\mathbf{A}) = 0$ if, and only if, $\mathbf{A} \in J$.

(5) Let $x_0, \ldots, x_{n-1}$ be a finite sequence in a Boolean algebra $\mathcal{B}$. We define $i(x) = m/n$, where $m$ is the largest integer $k \leqslant n$ such that $x_{i_0} \wedge \ldots \wedge x_{i_{k-1}} \neq 0$ for $i_0 < i_1 < \ldots < i_{k-1} < n$. If $A \subseteq \mathcal{B}$, we define the intersection number of $A$ by

$$i(A) = \inf \{ i(a) : a \in {}^{n}A, \text{ for some } n < \omega \} .$$

We say that $\mathcal{B}$ has the Kelley property (see Kelley 1959) if $\mathcal{B}$ -{0} is the countable union of sets with positive intersection number.

(6) If $I$ is a $\sigma$-ideal in $F$, $F/I$ is the quotient $\sigma$-Boolean algebra.

(7) We say that $\mathcal{B}$ is weakly $\sigma$-distributive if for every double sequence $x_{ij}$, $i,j<\omega$ of elements of $\mathcal{B}$ such that $x_{i,j+1} \leq x_{i,j}$ for every $i,j<\omega$, we have

$$\bigvee\{\bigwedge\{x_{ij} : j\in\omega\} : i\in\omega\} = \bigwedge\{\bigvee\{x_{i,\phi i} : i\in\omega\} : \phi\in {}^{\omega}\omega\},$$

where ${}^{\omega}\omega$ is the set of all functions from $\omega$ to $\omega$.

The theorem proved in Chuaqui 1977b, is the following:

Theorem. *Let $F$ be a $\sigma$-complete field of subsets of a set X, G a group of transformations of $F$, and $I \subseteq F$. Then the following conditions are necessary and jointly sufficient for the existence of a $\sigma$-additive, G-invariant, and $I$-positive probability measure on $F$ :*

(i) *$I$ is a $\sigma$-complete proper ideal of $F$;*

(ii) *all G-negligible sets belong to $I$;*

(iii) *If $A \in I$ and $A \approx_G B$, then $B \in I$;*

(iv) *$F/I$ has the Kelley property and is weakly $\sigma$-distributive.*

This theorem makes the existence of a measure dependent on the existence of a certain $\sigma$-ideal $I$, which is not very convenient in applications. The problem of obtaining conditions without this dependence remains open.

For the uniqueness of the measure, we have to add another condition (Chuaqui 198+). First, definitions.

(8) $A \approx_{G,I} B$, same as (6) with 'G-negligible' replaced by 'element of $I$'.

(9) $A \precsim_{G,I} B$ if and only if there is a $C\in F$ with

$$A \approx_{G,I} C \subseteq B.$$

Now, in order to insure the uniqueness of the measure, we have to add the condition:

(v) $A \precsim_{G,I} B$ *or* $B \precsim_{G,I} A$, *for all* $A, B \in F$.

Adding (v), (iv) can be omitted, and the measure $\mu$ is not only $G$-invariant but has the stronger property:

$$\mu(\mathbf{A}) < \mu(\mathbf{B}) \text{ if and only if } \mathbf{A} \lesssim_G \mathbf{B} .$$

Several other conditions equivalent to (v) are given in Chuaqui 198+.

The conditions for existence and uniqueness of the measure discussed up to now, depend on the $\sigma$-field of sets and the group of transformations. However, in our case, these are determined by a simple probability structure $K$. Thus, it would be very interesting to find conditions on $K$ itself. This is a completely open problem. It might be, for instance, that in terms of $K$ conditions for the uniqueness of the measure might be more natural, than that mentioned above.

Notice, however, that the fact of $\mathcal{B}_K$ and $G_K$ being obtained from $K$ does not insure the uniqueness of the measure. As an example, let us consider the following modification of Example 1. Instead of having the outcomes in $K_0$ of the form $\mathfrak{A}_S = \langle P, R_0, \ldots, R_{m-1}, S, a\rangle_{a \in P}$, we define a $K_5$ with outcomes $\overline{\mathfrak{A}}_S = \langle P, R, \ldots, R_{m-1}, S\rangle$ and intrinsic part $\langle P, R_0, \ldots, R_{m-1}\rangle$. The group $G_{K_5}$ is the same as $G_{K_0}$, but $\mathcal{B}_{K_5}$ is different from $\mathcal{B}_{K_0}$. The atoms of $\mathbf{B}_{K_5}$ are the sets of models of sentences of the form 'there exist exactly $p$ elements in $\bigcap_{i<m+1} T_i$' where $T_i$ is $R_i$ or $\sim R_i$, and $T_m$ is $S$ or $\sim S$. (This can be proved by elimination of quantifiers.) Hence we do not have equivalent atoms, and so there are many invariant measures.

### 7.2 Conditions of a logical character for the existence and uniqueness of measures

The simple probability structure $K = \langle \mathbf{K}, \mathfrak{B}\rangle$ may be given by a theory in a language $\mathcal{L}$; i.e. $\mathbf{K}$ may be the set of all models of a given theory (with a given universe). It would be of interest to analyze conditions on the theory or the language for the existence and uniqueness of measures. There has been much done, especially in the field of Model Theory called Stability Theory, on isomorphisms of relational systems. These results may be applicable in our case. However, this is a field that is completely unexplored.

### 7.3 Isomorphisms of cylindric algebras

Let $K = \langle \mathbf{K}, \mathfrak{B}\rangle$ be a simple probability structure where the elements of $\mathbf{K}$ are of the form $\mathfrak{A} = \langle A, R_i^{\mathfrak{A}}\rangle_{i \in I^{\mathfrak{A}}}$. Let $f \in G_K$; I shall obtain an isomorphism $\overline{f}$ of $\mathfrak{C}_{\mathbf{K}}$ derived from f.

(i) For $i \in I^{\mathfrak{B}}$, let

$$\bar{\delta}(\langle R_i^{\mathfrak{a}} : \mathfrak{a} \in K \rangle) = \langle R_i^{\mathfrak{a}} : \mathfrak{a} \in K \rangle$$

For $i \in I^{\mathfrak{a}} \sim I^{\mathfrak{B}}$, let

$$\bar{\delta}(\langle R_i^{\mathfrak{a}} : \mathfrak{a} \in K \rangle) = \langle \delta(R_i^{\mathfrak{a}}) : \mathfrak{a} \in K \rangle .$$

(ii) $\bar{\delta}(d_{\kappa\lambda}) = d_{\kappa\lambda}$

(iii) $\bar{\delta}(A + B) = \bar{\delta}(A) + \bar{\delta}(B)$, $\bar{\delta}(A \cdot B) = \bar{\delta}(A) \cdot \bar{\delta}(B)$

(iv) $\bar{\delta}(\bigwedge \{A_i : i \in \omega\}) = \bigwedge \{\bar{\delta}(A_i) : i \in \omega\}$

$\bar{\delta}(\bigvee \{A_i : i \in \omega\}) = \bigvee \{\bar{\delta}(A_i) : i \in \dot{\omega}\}$

(v) $\bar{\delta}(c_\kappa X) = c_\kappa \bar{\delta}(X)$

$\bar{\delta}$ is an isomorphism of $\mathfrak{C}_K$ onto another cylindric algebra $\overline{\mathfrak{C}}$

The condition we imposed on $\delta$ in order to belong to $G_K$, can now be restated simply:

For any 0-dimensional element $X \in C_K$, $\bar{\delta}(X) \in C_K$.

Thus the study of certain isomorphisms of cylindric algebras might help to solve our problems, especially those questions about the existence and unicity of measures. This is a possible new avenue of study.

## 8. APPLICATIONS

There are two main areas of application of my definition of probability: statistical inference and physical theory.

### 8.1 Statistical inference

Using my approach to probability, especially the compound probability structures which have not been defined in this paper, I have constructed models for the two main types of statistical inference: classical and Bayesian. (See Chuaqui 1980 and 1982). As is well-known, these are, on the main, competing schools. (See Barnett 1973.) It would be of interest to study concrete examples, for instance in economical science, where both methods can be used. Models for both according to my definition would be constructed, and, then, compared. It may be possible to discover advantages and disadvantages of each method from a theoretical point of view.

For certain statistical methods, there is still need of developing models according to my definition. As a matter of fact, I do not yet know whether models are possible for some of them (for instance, fiducial methods).

In any case, the models developed in Chuaqui 1980 and 1982 for classical and Bayesian inference, are just outlined there and much detailed work needs to be done.

## 8.2 Physical Theory

Probability occurs in physical theories such as statistical mechanics and quantum mechanics. I believe that the construction of my models for the first of these, presents no essential problem. However, the situation is different for quantum mechanics. Here, it might be necessary to modify the models. However, if the attempt to explain probability as used in Quantum Mechanics succeeds, it may lead to a better formulation of the theory.

Another area related to physical theory, where progress has been made, is in the formulation of stochastic processes. Here, Bertossi 1982 has obtained models for random walks and Brownian motion.

## REFERENCES

V.Barnett

1973 Comparative Statistical Inference, John Wiley and Sons Ltd., London.

L.Bertossi

1982 *Chuaqui's definition of probability in some stochastic processes*, to appear in the Proceedings of the V Latin American Symposium on Mathematical Logic, to be published by Marcel Dekker, New York.

R.Carnap

1971 *A basic system of inductive logic*, Part I, in Studies in Inductive Logic and Probability I, R.Carnap and R.C. Jeffrey (editors), U. of California Press, Berkeley and Los Angeles, pp. 33 - 165.

R.Chuaqui

1965 *A definition of probability based on equal likehood*, Ph.D. Dissertation, U. of California, Berkeley.

1975 *A model-theoretical definition of probability*, Contributed Papers, 5th International Congress of Logic, Methodology, and Philosophy of Science, London, Ontario, Canadá, pp. VI, 7-8.

1977a *A semantical definition of probability*, in Non-Classical Logics, Model Theory and Computability, Arruda, da Costa and Chuaqui (eds.), North-Holland Pub. Co., Amsterdam, pp. 135-167.

1977b *Measures invariant under a group of transformations*, Pac.J. of Math., vol. 68, pp. 313-328.

1979 *Probability as between Truth and Falsehood*, Abstracts, Sections 5-7, 6th International Congress of Logic, Methodology and Philosophy of Science Hanover, 1979, pp. 227-231.

1980 *Foundations of statistical methods using a semantical definition of probability*, in Mathematical Logic in Latin America, Arruda, Chuaqui, da Costa (eds.), North-Holland Pub. Co., Amsterdam, pp. 103-120.

1982 *Factual and cognitive probability*, to appear in the Proceedings of the V Latin American Symposium on Mathematical Logic, to be published by Marcel Dekker, New York.

198+ *Simple cardinal algebras and their applications to invariant measures*, to appear.

H.Enderton

1972 A Mathematical Introduction to Logic, Academic Press, New York.

K. Gödel

1939 *Consistency proof for the generalized continuum hypothesis*, Proc. Nat. Acad. Sci., vol. 25, pp. 220-224

L.Henkin, J.D.Monk and A.Tarski

1981 Cylindric Set Algebras, Lecture Notes in Mathematics 883, Springer-Verlag, Berlin,Heidelberg, New York, pp. 1-129.

H. J. Keisler

1971 Model Theory for Infinitary Logic, North-Holland Pub.Co., Amsterdam.

J.L.Kelley

1959 *Measures on Boolean algebras*, Pac. J. of Math., vol. 9, 1165-1178.

K.Kuratowski and A.Mostowski

1976 Set theory, with an introduction to Descriptive Set Theory, 2nd Edition, North-Holland Pub.Co., Amsterdam.

B.Mates

1964 Elementary Logic, Oxford U. Press, New York.

Y.N. Moschovakis

1980 Descriptive Set Theory, North-Holland Pub. Co., Amsterdam.

R.M. Solovay

1970 *A model of set theory in which every set is Lebesgue measurable*, Annals of Math., vol. 92, pp. 1-56.

A.Tarski

1935 *Der Wahrheitsbegriff in den formalisierten Sprachen*, Studia Philos. (Warsaw) vol. 1 pp. 261-405. (English trans. in Logic, Semantics and Metamathematics, Oxford U. Press, pp. 152-278.)

1949 Cardinal Algebras, Oxford U. Press.

1969 *Truth and Proof*, Scientific American, vol. 220, pp. 63-77.

# 8
# Sur la Construction des Représentations des Groupes Classiques

JORGE SOTO-ANDRADE / Departamento de Matemáticas, Facultad de Ciencias, Básicas y Farmacéuticas, Universidad de Chile, Santiago, Chile

L'un des problèmes fondamentaux de la théorie de représentations des groupes est la construction explicite des représentations unitaires irréductibles d'un groupe G donné. Dans cet exposé, nous nous intéréssons au cas aù G est un groupe classique, défini sur un corps local ou fini. Après un bref aperçu historique des méthodes de construction connues, nous présentons quelques résultats partiels et signalons des questions ouvertes concernant spécialement la construction de "représentations de Weil généralisées".

## § 1. Méthodes connues

### 1. La classification des représentations des groupes classiques

Déjà dans les années 50, il est apparu dans les travaux de l'école soviétique (I.M. GELFAND et ses collaborateurs) sur les groupes classiques semi-simples réels et complexes, que les représentations unitaires irréductibles de ces groupes devraient s'organiser en séries, chacune associée de manière naturelle à une classe de conjugaison de sous-groupes de Cartan du groupe consideré. Rappellons que, pour un tel groupe G , les sous-groupes de Cartan sont les sous-groupes commutatifs de G formés d'éléments diagonalisables sur $\mathbb{C}$ , maximaux par rapport à ces deux propriétés. Ainsi par exemple, pour $G = SL(2,\mathbb{R})$ , nous avons les deux représentants suivants pour les (classes de conjugaison des) sous groupes de Cartan

$$T_1 = \left\{ \begin{pmatrix} r & 0 \\ 0 & r^{-1} \end{pmatrix} \mid r \in \mathbb{R}^\times \right\} \simeq \mathbb{R}^\times$$

$$T_2 = \left\{ \begin{pmatrix} \cos\theta & -\sin\theta \\ \sin\theta & \cos\theta \end{pmatrix} \middle| 0 \leq \theta < 2\pi \right\} \simeq U$$

(où U désigne le groupe du cercle unité dans $\mathbb{C}$). La série principale de représentations de G est associée à $T_1$ et paramétrée par les caractères de $\mathbb{R}^\times$ . La série discrète de représentations de G est associée à $T_2$ et paramétrée par $\mathbb{Z}^+$ , le groupe de caractères de U . Ces deux séries fournissent "essentiellement toutes" les représentations unitaires irréductibles de G , c'est-à-dire assez pour établir la formule de Plancherel. Ceci reste en fait vrai pour un corps local ou fini; dans ce dernier cas, on a même exactement toutes les représentations (unitaires) irréductibles de G .

La classification des représentations unitaires irréductibles des groupes de Lie semi-simples réels suivant ce schéma a été accomplie dans les années 60 par HARISH-CHANDRA (cf. [H-C 1] , étendue ensuite au cas p-adique (cf. [H-C 2]) et même au cas fini (cf. [H-C 3] ). Cela ne résout cependant pas le problème de la construction explicite de ces représentations, tout particulièrement celles de la série discrète de G .

Rappelons que dans le cas réel (ou p-adique) la série discrète consiste, par définition, des représentations unitaires irréductibles de G à fonctions coefficients matriciels de carré intégrable sur G , c'est-à-dire, de celles qui aparaîssent discrètement dans la formule de Plancherel. Dans le cas fini, on appelle en fait série discrète, la famille des représentations (super)-cuspidales de G , c'est-à-dire celles qui n'admettent pas de vecteur fixe par aucun radical unipotent d'un sous-groupe parabolique (c'est-à-dire contenant un sous-groupe de Borel) de $G$ . Pour $G = SL(2,k)$ , cela signifie simplement ne pas avoir de vecteur fixe par aucun conjugué du sous-groupe des matrices unipotentes supérieures de $G$ .

HARISH-CHANDRA a néanmoins montré que $G$ possède une série discrète si et seulement si $G$ a un sous-groupe de Cartan compact et décrit alors les caractères de ces représentations.

## 2. Construction de représentations à la BOREL-WEIL-BOTT

Dans le cas d'un groupe de Lie compact $G$ et d'un tore maximal $T$ de $G$ (unique à conjugaison près), on sait que l'espace quotient $G/T$ peut être muni d'une structure de variété complexe G-homogène. Alors chaque caractère $\omega$ de $T$ fournit un fibré en droites holomorphe G-homogène $L_\omega$ sur $G/T$ , qui est caracterisé par le fait que $T$ agit dans la fibre au-dessus de $eT$ via $\omega$ . L'action naturelle de $G$ dans $L_\omega$ passe aux groupes de cohomologie

$$H^k(G/T, \mathcal{O}(L_\omega))$$

du faisceau $\mathcal{O}(L_\omega)$ des germes de sections holomorphes de $L_\omega$ . Ces groupes de cohomologie deviennent alors des représentations de $G$ , qui sont de dimension finie car $G/T$ est compact. D'après la théorie de Hodge, ces espaces sont formés des $(0,k)$-formes harmoniques à valeurs dans $L_\omega$ .

Le théorème de BOREL-WEIL-BOTT nous dit alors que, génériquement en $\omega$ , ces groupes de cohomologie sont non nuls pour un seul entier $k = k(\omega)$ , cas où l'on obtient une représentation irréductible de $G$ . On peut réaliser ainsi toute représentation (unitaire) irréductible de $G$ , même avec $k(\omega) = 0$ , cas où le groupe de cohomologie correspondant se réduit à l'espace des sections holomorphes du fibré en droites $L_\omega$ . On obtient ainsi une réalisation "géométrique" des représentations unitaires irréductibles de $G$ .

Dans le cas d'un groupe de Lie (connexe) semi-simple réel $G$ , on peut essayer d'étendre la construction précédente pour obtenir une construction de la série discrète. Si $G$ a une série discrète, $G$ possède un sous-groupe de Cartan compact, qui est en particulier un tore maximal. Comme dans le cas compact, on peut fabriquer, étant donné un caractère $\omega$ de $T$ , un fibré en droites holomorphe G-homogène $L_\omega$ sur la variété complexe G-homogène $G/T$ (qui sera non-compacte dès que $G$ est non compact). Par exemple, pour $G = SL(2,\mathbb{R})$ , on a $T = SO(2,\mathbb{R})$ et $G/T$ n'est autre que le demi-plan superieur de Poincaré. Dans ce cas, les espaces $\mathcal{H}(L_\omega)$ des sections holomorphes de carré intégrable de $L_\omega$ sont encore, de manière naturelle, des espaces de représentation de $G$ , mais ne fournissent qu'une petite partie de la série discrète de $G$ . Pour obtenir toute la série discrète, il faut, en suivant une suggestion de R.P. LANGLANDS, considérer, plus généralement, des groupes de cohomologie de carré integrable $\mathcal{H}^k(L_\omega)$ , c'est-à-dire les espaces de $(0,k)$-formes sur $G/T$ à valeurs dans $L_\omega$ qui sont harmoniques et de carré intégrable. On a ainsi, de manière naturelle, des représentations unitaires de $G$ , qui ne sont plus des groupes de cohomologie faisceautique comme dans le cas compact. W.SCHMID a demontré [cf. S1] que, génériquement en $\omega$ , ces représentations $\mathcal{H}^k(L_\omega)$ sont non-nulles seulement pour un entier $k$ dépendant de $\omega$ , et fournissent alors une représentation irréductible qui appartient à la série discrète de $G$ . De plus, on obtient ainsi des modèles pour toute la série discrète de $G$ . (voir aussi [A-S] ).

Dans le cas d'un corps fini, DELIGNE et LUSZTIG ([D-L] ) ont réussi, en 1976, à obtenir des modèles pour "presque toutes" les représentations irréductibles complexes d'un groupe classique (plus précisément un groupe algébrique réductif sur un corps fini $\mathbb{F}_q$) avec une construction à la BOREL-WEIL-BOTT, où les espaces $\mathcal{H}^k(L_\omega)$ sont remplacés par les groupes de cohomologie étale $\ell$-adique de certaines variétés algébriques sur la clotûre algébrique de $\mathbb{F}_q$ , sur lesquelles le groupe agit de manière naturelle.

(Rappelons que la clotûre algébrique du corps des nombres $\ell$-adiques, où $\ell$ est un nombre premier, est isomorphe à $\mathbb{C}$ ).

## 3. Les représentations de Weil

Une approche fort différente au problème de construction de représentations irreductibles (unitaires) des groupes classiques est suggerée par les représentations de Weil des groupes symplectiques (cf. [W], [L-V]). Nous esquissons ci-dessous une manière "non classique" d'obtenir ces représentations (d'après une remarque de P. CARTIER). Pour simplifier l'exposé, nous nous bornons au cas de $G = SL(2,k)$, où $k$ est un corps fini $\mathbb{F}_q$ .

Le groupe $G$ agit de manière naturelle, par exemple par multiplication matricielle à droite, sur le plan fini $k^2$ . On en déduit une représentation unitaire $(V,\tau)$ de $G$ , où $V = L^2(k^2)$ , et dont l'action $\tau$ est donnée par

$$(\tau_g f)(x) = f(xg) \qquad (g \in G , \quad f \in V , \quad x \in k^2).$$

Par décomposition, suivant le groupe de caractères $(k^\times)^\wedge$ de $k^\times$ , cette représentation fournit la série principale de $G$ . On a ainsi une construction géométrique de cette série. Par contre, la série discrète de $G$ n'est pas accessible à une construction géométrique aussi simple. Néanmoins, on peut transformer la représentation $\tau$ de manière que sa nouvelle version suggère comment construire la série discrète de $G$ .

On fixe un caractère additif non trivial, noté $e$ , de $k^+$ . On définit alors une transformée de Fourier partielle $F$ , relative à la seconde variable, notée aussi $f \mapsto \hat{f}$ , de $V = L^2(k^2)$ dans lui-même, par

$$\hat{f}(x) = q^{-1/2} \sum_{y_2 \in k} e(-x_2 y_2) f(x_1, y_2)$$

pour $f \in V$ et $x \in k^2$ . Si pour chaque $g \in G$ , on note $\hat{\tau}_g$ le transformé $F \circ \tau_g \circ F^{-1}$ de $\tau_g$ suivant $F$ , on trouve sur l'ensemble de générateurs de $G$ formé des éléments

$$u(s) = \begin{pmatrix} 1 & s \\ 0 & 1 \end{pmatrix} \qquad (s \in k) ,$$

$$h(t) = \begin{pmatrix} t & 0 \\ 0 & t^{-1} \end{pmatrix} \qquad (t \in k^{\times}) , \quad w = \begin{pmatrix} 0 & 1 \\ -1 & 0 \end{pmatrix}$$

les formules suivantes (pour $f \in V$ et $x \in k^2$)

(1) $(\hat{\tau}_{u(s)} f)(x) = e(sx_1x_2)f(x)$ $\qquad (s \in k)$ ,

(2) $(\hat{\tau}_{h(t)} f)(x) = f(tx)$ $\qquad (t \in k^{\times})$ ,

(3) $(\hat{\tau}_{w} f)(x) = q^{-1} \sum_{y \in k^2} e(x_1y_2 + x_2y_1)f(y).$

On voit ainsi que les opérateurs $\hat{\tau}_g$ $(g \in G)$ s'expriment en fait en termes de la forme quadratique hyperbolique $H : x \mapsto x_1x_2$ $(x \in k^2)$ et de sa forme bilinéaire associée $B_H$ définie par

$$B_H(x,y) = H(x + y) - H(x) - H(y) \qquad (x,y \in k^2) .$$

Cela suggère de construire une représentation de $G$ associée à l'autre plan quadratique non-dégénéré sur $k$ , à savoir celui défini par la norme $N$ de l'unique extension quadratique $K$ de $k$ , en remplaçant partout $x \in k^2$ par $z \in K$ , la forme $H$ par $N$ et, bien entendu $B_H$ par $B_N$ , qui est donnée par

$$B_N(z,z') = Tr(z\bar{z}') \qquad (z,z' \in K) ,$$

où $Tr$ désigne la trace de $K$ sur $k$ . On obtient en fait ainsi une représentation de $G$ , quitte à changer le signe de la constante $q^{-1}$ dans la formule (3). Plus généralement, pour un espace quadratique non-dégénéré $(E,Q)$ de dimension <u>paire</u> sur $k$ , on obtient une représentation de $G$ à l'aide des formules (1) à (3), pour $f \in V$ , $x \in E$ , quitte à remplacer $H$ par $Q$ (et donc $B_H$ par $B_Q$), ainsi que la constante $q^{-1}$ dans (3) par $\varepsilon(Q)q^{-1}$ , où le signe $\varepsilon(Q)$ de $Q$ est, par définition, égal à 1 si $Q$ est déployée (c'est-à-dire, somme orthogonale de plans hyperboliques) et à -1 sinon. La démostration se fait en vérifiant que les relations entre les générateurs donnés sont respectés. La seule relation non triviale est

$$wu(t)w = h(-t^{-1})u(-t)wu(-t^{-1}) \qquad (t \in k^{\times}) ,$$

où la vérification se ramène au calcul (immédiat) de la somme de Gauss quadratique associée à $t^{-1}Q$ ,

$$\sum_{x \in E} e(t^{-1}Q(x)) = \varepsilon(Q)q \qquad (t \in k^{\times})$$

(cf.[SA1]) .

De cette manière, on retrouve, dans notre cas, ce qu'on appelle la représentation de Weil associée a l'espace quadratique $(E,Q)$ . Le point le plus intéressant pour nous ici, est que pour $(E,Q) = (K,N)$ , cette représentation fournit, par décomposition, suivant le groupe orthogonal $O(Q)$ de $Q$ , la série discrète de représentations de $G$ .

## § 2. Méthodes "non-classiques"

### 1. Introduction

Les résultats précédents suggèrent de chercher une construction analogue pour la série discrète, ainsi que les autres séries de représentations irréductibles, du groupe $SL(3,k)$ et plus géneralement $SL(n,k)$ . D'autre part, d'après une remarque de P. CARTIER, il n'est pas difficile de construire une représentation de Weil pour $GL(2,k)$ (il s'agit en fait de l'induite à $GL(2,k)$ de la représentation de Weil de $SL(2,k)$), dont l'étude est bien plus aisée que celle de la représentation de Weil de $SL(2,k)$ . Le problème est donc aussi ouvert de construire des représentations de Weil généralisées pour les groupes $GL(n,k)$ , qui seraient associées à des n-formes el fourniraient, dans les cas des normes des différentes algèbres semi-simples de dimension $n$ sur $k$ , les différentes séries de représentations irréductibles du groupe $GL(n,k)$ . Tout particulièrement la série discrète devrait être fournie par la représentation de Weil généralisée associée à l'unique extension de degré $n$ du corps fini $k$ . De plus, comme il est le cas pour $GL(2,k)$ et $SL(2,k)$ , on espère avoir une construction si universelle, qu'elle se transpose sans peine au cas de corps locaux et fournisse "essentiellement toutes" les représentations unitaires irréductibles dans ce cas.

Nous présentons ci-dessous, une construction d'une telle représentation, dans le cas cubique de $G = GL(3,k)$ . Nous montrons en fait comment passer des modèles de Kirillov de la série discrète de $G$ (que nous construisons de manière adaptée à notre but) à cette représentation de Weil généralisée. Après quelques préliminaires, nous commençons par reprendre de ce point de vue le cas des représentations de Weil quadratiques ($n = 2$) .

## 2. Préliminaires: le lemme de Silberger

Nous nous servirons dans notre construction du

LEMME 1 (Silberger). *Soit* $(V,\sigma)$ *une représentation irreductible d'un groupe fini* $H$ . *Alors tout endomorphisme* $A$ *de l'espace vectoriel complexe* $V$ *peut s'exprimer sous la forme*

$$A = \frac{\dim V}{|H|} \sum_{h\in H} T_r(A\sigma_h^{-1})\sigma_h \ . \qquad (1)$$

*Démonstration*: La représentation $\sigma$ étant irréductible, on sait bien que $A$ peut s'exprimer comme combinaison linéaire des opérateurs $\sigma_h$ $(h \in H)$. Il suffit donc de vérifier le lemme pour $A = \sigma_{h_o}$ , pour $h_o \in H$ arbitraire. Mais, par composition avec $\sigma_{h_o}^{-1}$ , on se ramène au cas $A = Id$ , où le lemme exprime simplement le fait que le projecteur isotypique de $(V,\sigma)$ associé au type d'isomorphie de $\sigma$ est l'identité de $V$ .

C.Q.F.D.

Nous emploierons en fait ce lemme dans le cas particulier du corollaire 2 ci-dessous.

COROLLAIRE 1. *Soit* $(V,\pi)$ *une représentation irréductible d'un groupe fini* $G$ *dont la restriction* $\sigma$ *à un certain sous-groupe* $H$ *de* $G$ *est encore irréductible. Si* $\chi$ *désigne le caractère de* $\pi$ *, on a alors*

$$\pi_g = \frac{\dim V}{|H|} \sum_{h\in H} \bar{\chi}(hg^{-1})\sigma_h \qquad (g \in G). \tag{2}$$

COROLLAIRE 2. *Soit* $(V,\pi)$ *une représentation irréductible d'un groupe fini* $G$ *dont la restriction à un certain sous-groupe* $H$ *de* $G$ *est encore irréductible et coïncide, en plus, avec l'induite à* $H$ *, d'une représentation unidimensionelle* $\gamma$ *d'un sous groupe* $U$ *de* $H$ *. Notons* $\chi$ *le caractère de* $\pi$ *. Alors, pour chaque* $g \in G$ *, on a*

$$(\pi_g f)(h) = \sum_{h'\in U\backslash K} K_g(h,h')f(h') \qquad (f \in V\ ,\quad h \in H)\ , \tag{3}$$

*où le noyau* U-*bihomogène* $K_g$ *de l'opérateur* $\pi_g$ *(qui vérifie donc la relation* $K_g(uh,u'h') = \gamma(u)K_g(h,h')\bar{\gamma}(u')$ *, quels que soient* $h,h' \in H$ *,* $u,u' \in U$*) est donné par*

$$K_g(h,h') = \frac{1}{|U|} \sum_{u\in U} \gamma(u)\bar{\chi}(h^{-1}uh'g^{-1}) \qquad (h,h' \in H)\ . \tag{4}$$

Ces corollaires sont des conséquences immédiates du lemme 1.

## 3. *Le cas quadratique*: $G = GL(2,k)$

### 3.1.

Rappelons que l'on note $k$ le corps fini à $q$ éléments et $K$ son unique extension quadratique. On désigne par $N$ (resp. $T_r$) la norme (resp. trace) de $K$ sur $k$

Nous montrons tout d'abord, comment on peut récuperer la représentation de Weil associée au plan quadratique $(K,N)$ à partir des modèles de Kirillov pour les représentations de la série discrète de $G$ .

Désignons par $\pi^\Lambda$ $(\Lambda \in (K^\times)^\wedge - (k^\times)^\wedge$ les représentations de cette série. Rappelons que $\dim \pi^\Lambda = q - 1$ pour tout $\Lambda$ , et que $\pi^\Lambda \simeq \pi^{\Lambda'}$ si et seulement si $\Lambda' = \Lambda$ ou $\Lambda' = \Lambda^q$ .

3.2. Les modèles de Kirillov pour la série discrète de $G$

Chaque représentation $\pi^\Lambda$ de la série discrète de $G$ a un modèle de Kirillov dans l'espace $L^2(k^\times) = \mathbb{C}^{k^\times}$, que l'on peut construire comme suit. On note $H$ le sous-groupe de $G$ formé des

$$h(r,s) = \begin{pmatrix} r & s \\ 0 & 1 \end{pmatrix} \qquad (r \in k^\times , \quad s \in k) .$$

On choisit un caractère non-trivial $e$ de $k^+$ et l'on forme la représentation $\sigma$ de $H$ induite par la représentation $\underline{e} : h(1,s) \longmapsto e(s)$ du sous-groupe $U$ de $H$ formé des $h(1,s)$ $(s \in k)$. L'espace de $\sigma$ s'identifie a $L^2(k^\times)$ et son action est alors donnée par la formule

$$[\sigma_{h(r,s)} f](t) = e(ts)f(rt) \tag{5}$$

pour $f \in L^2(k^\times)$, $r,t \in k^\times$, $s \in k$. Cette représentation est irréductible.

Dans le modèle de Kirillov de $\Pi^\Lambda$ dans $L^2(k^\times)$, le sous-groupe $H$ agit suivant $\sigma$ et le centre de $G$ agit suivant la restriction de $\Lambda$ à $k^\times$. Pour achever la description du modèle, il suffit de donner l'action de $w = \begin{pmatrix} 0 & 1 \\ -1 & 0 \end{pmatrix}$, ce que nous faisons à l'aide du corollaire 2, à partir du caractère $\chi^\Lambda$ de $\Pi^\Lambda$.

On trouve, d'après ce corollaire, avec $g = w$ et $K_g = W$, en identifiant $h(t,0)$ et $t$ pour $t \in k^\times$,

$$W(r,r') = \frac{1}{q} \sum_{s \in k^+} e(s)\overline{\chi^\Lambda}(h(r^{-1},0)h(1,s)h(r',0)w) .$$

Or, rappelons que les valeurs du caractère $\chi^\Lambda$ sur les classes de conjugaison non-scalaires de $G$ sont données par

$$\chi^\Lambda(g) = - \sum_{z \in \phi_K^{-1}(\phi_G(g))} \Lambda(z) \qquad (g \in G \quad Z(G)) , \tag{6}$$

où l'on a posé

$$\phi_G(g) = (\det g, \operatorname{tr} g) \qquad (g \in G) ,$$

$$\phi_K(z) = (Nz, \operatorname{Tr} z) \qquad (z \in K) .$$

On en tire

$$\overline{\chi}^{\Lambda}(h(r^{-1},0)h(1,s)h(r',0)w) = \overline{\chi}^{\Lambda}\begin{pmatrix} r^{-1}s & -r'r^{-1} \\ 1 & 0 \end{pmatrix}$$

$$= - \sum_{\substack{\operatorname{Tr} z = r^{-1}s \\ Nz = r'r^{-1}}} \overline{\Lambda}(z) ,$$

d'où

$$W(r,r') = \frac{-1}{q} \sum_{z \in K} e(r\operatorname{Tr} z)\overline{\Lambda}(z)\delta_{r',rNz} \qquad (r,r' \in k^{\times}) , \quad (7)$$

c'est-à-dire

$$(\pi_w^{\Lambda} f)(r) = - \frac{1}{q} \sum_{z \in K} e(r\operatorname{Tr} z)\overline{\Lambda}(z)f(rNz) , \qquad (8)$$

quels que soient $f \in L^2(k^{\times})$ , $r \in k^{\times}$ .

### 3.3. Réalisation des modèles de Kirillov dans $L^2(K^{\times} \times k^{\times})$

Les modèles de Kirillov apparaîssent comme des composantes d'une représentation de $G$ dans $L^2(K^{\times} \times k^{\times})$ , en associant (pour $\Lambda \in (K^{\times})^{\wedge} - (k^{\times})^{\wedge}$) à chaque $f \in L^2(k^{\times})$ , la fonction, notée provisoirement $\widetilde{f}$ , de $K^{\times} \times k^{\times}$ dans $\mathbb{C}$ , définie par

$$\widetilde{f}(z,t) = \overline{\Lambda}(z)f(tN(z)) \qquad (z \in K^{\times} , t \in k^{\times}) . \quad (9)$$

PROPOSITION 1. <u>L'application</u> $f \mapsto \widetilde{f}$ <u>ainsi définie, est un isomorphisme entre notre modèle de Kirillov</u> $(L^2(k^{\times}),\pi^{\Lambda})$ <u>et la représentation</u> $(V_{\Lambda},\rho^{\Lambda})$ <u>de</u> $G$ , <u>dont l'espace est formé des fonctions</u> $\widetilde{f} \in L^2(K^{\times} \times k^{\times})$ <u>vérifiant</u>

$$\widetilde{f}(z'z, tN(z')^{-1}) = \overline{\Lambda}(z')\widetilde{f}(z,t) \qquad (z,z' \in K^{\times} , t \in k^{\times}) ,$$

et dont l'action $\rho^{\Lambda}$ est donnée par les formules

$$(\rho^{\Lambda}\begin{pmatrix}1 & s\\ 0 & 1\end{pmatrix}\tilde{f})(z,t) = e(stN(z))\tilde{f}(z,t) \qquad (s \in k) , \qquad (10)$$

$$(\rho^{\Lambda}\begin{pmatrix}r & 0\\ 0 & r^{-1}\end{pmatrix}\tilde{f})(z,t) = \tilde{f}(rz,t) \qquad (r \in k^{\times}) , \qquad (11)$$

$$(\rho^{\Lambda}\begin{pmatrix}r & 0\\ 0 & 1\end{pmatrix}\tilde{f})(z,t) = \tilde{f}(z,rt) \qquad (r \in k^{\times}) , \qquad (12)$$

$$(\rho^{\Lambda}\begin{pmatrix}0 & 1\\ -1 & 0\end{pmatrix}\tilde{f})(z,t) = -q^{-1}\sum_{y\in K^{\times}} e(t\ Tr(z\bar{y}))\tilde{f}(y,t) , \qquad (13)$$

pour $\tilde{f} \in V_{\Lambda}$ , $z \in K^{\times}$ , $t \in k^{\times}$ .

Démonstration: Les formules (10), (11), (12) résultent aussitôt des formules (5), (8) et (9). Enfin (11) résulte aussitôt de la décomposition

$$\begin{pmatrix}r & 0\\ 0 & r^{-1}\end{pmatrix} = \begin{pmatrix}r^{-1} & 0\\ 0 & r^{-1}\end{pmatrix}\begin{pmatrix}r^{2} & 0\\ 0 & 1\end{pmatrix} \qquad (r \in k^{\times}) ,$$

de (9) pour $z = r$ , et du fait que le centre de $G$ agit suivant $\Lambda$ dans le modèle de Kirillov $\pi^{\Lambda}$ .

3.4. La représentation de Weil de $G$ associée à $(K,N)$

Nous avons obtenu des réalisations $(V_{\Lambda},\rho^{\Lambda})$ des modèles de Kirillov de la série discrète de $G$ comme des composantes de la représentation $(V',\rho')$ de $G$ , dont l'espace $V'$ est formé des fonctions $f'$ de $K^{\times} \times k^{\times}$ dans $\mathbb{C}$ vérifiant la condition

$$\sum_{Nz=r} f'(z,t) = 0 \qquad (r,t \in k^{\times}) ,$$

et dont l'action $\rho'$ est donnée par les formules (10) à (13) ci-dessus, pour $\rho'$ à la place de $\rho^{\Lambda}$ et $f'$ à la place de $\tilde{f}$ .

En prolongeant cette représentation à l'espace $L^{2}(K^{\times} \times k^{\times})$ tout entier, par les mêmes formules, on retrouve enfin la représentation de Weil associée au plan quadratique $(K,N)$ .

## 4. Le cas cubique: $G = GL(3,k)$

### 4.1.

Nous désignerons par $\mathbb{K}$ l'unique extension cubique du corps fini $k$ et par $\mathbb{N}$ (resp. $\mathbb{T}r$) la norme (resp. trace) de $\mathbb{K}$ sur $k$.

Nous procédons de manière analogue au cas quadratique, en réalisant les modèles de Kirillov de la série discrète de $G$ comme des composantes d'une représentation $(V',\rho')$ de $G$, qui apparaîtra alors comme une "représentation de Weil généralisée" associée à l'espace cubique $(\mathbb{K},\mathbb{N})$.

Dorénavant $GL(2,k)$ sera noté $G_2$. De même le sous-groupe $H$ (resp. $U$) du numéro précédent sera noté $H_2$ (resp. $U_2$). Les éléments $u(s)$ $(s \in k)$ de $U_2$ seront notés désormais $u_2(s)$ $(s \in k)$. On notera $B_2$ le sous-groupe des matrices triangulaires supérieures de $G_2$.

### 4.2. La représentation distinguée $\tau$ du sous groupe $H$ de $G$

On note $H$ le sous-groupe de $G$ formé des matrices de la forme

$$h(a,b) = \begin{pmatrix} a & b \\ 0 & 1 \end{pmatrix} \qquad (a \in G_2 , \quad b \in k^2) .$$

Les représentations de la série discrète de $G$ ont une restriction commune à $H$, notée $\tau$, qui est encore irreductible et de dimension $(q^2 - 1)(q - 1)$. Nous appelons cette représentation $\tau$, qui est de dimension maximale parmi les représentations de $H$, représentation distinguée de $H$.

Notons $U$ le sous-groupe des matrices unipotentes supérieures de $G$. Fixons un caractère non-trivial $e$ de $k^+$ et notons $\underline{e}$ le caractère de $U$ défini par

$$\underline{e}(u) = e(u_{12} + u_{23}) \qquad (u = (u_{ij})_{ij} \in U) .$$

La représentation $\tau$ n'est autre que l'induite par la représentation unidimensionelle $\underline{e}$ de $U$. Elle peut être réalisée dans l'espace $M$ de toutes les fonctions $f$ de $G_2$ dans $\mathbb{C}$ telles que

$$f(u_2(s)c) = e(s)f(c) \qquad (s \in k\ ,\ c \in G_2)\ . \tag{14}$$

L'action $\tau$ est alors donnée par

$$[\tau_{h(a,b)}f](c) = e((c_{21},c_{22})\cdot b)f(ca)\ , \tag{15}$$

quels que soient $a,c \in G_2$ , $b \in k^2$ , $f \in M$ .

4.3. Les modèles de Kirillov pour la série discrète de $G$

4.3.1.

Soit $\Omega \in (\mathbb{K}^\times)^\wedge \smallsetminus (k^\times)^\wedge$ (c'est-à-dire, on suppose que $\Omega$ n'est pas de la forme $\alpha \circ \mathbb{N}$ , pour $\alpha \in (k^\times)^\wedge$ , autrement dit $\Omega \neq \Omega^q$). Pour décrire un modèle de Kirillov de la représentation correspondante $\pi^\Omega$ de la série discrète de $G$ , il suffit d'étendre la représentation $\tau$ de de $H$ dans $M$ à tout entier, en décrivant, par exemple, l'opérateur $W_{23}^\Omega := \pi^\Omega_{w_{23}}$ correspondant à l'élement de Weyl

$$w_{23} = \begin{pmatrix} 1 & 0 & 0 \\ 0 & 0 & 1 \\ 0 & -1 & 0 \end{pmatrix}\ ,$$

l'action du centre de $G$ étant donnée par la restriction de $\Omega$ à $k^\times$ .

4.3.2. Forme générale des opérateurs $W_{23}^\Omega$

L'opérateur $W_{23}^\Omega$ se calcule sans difficulté à l'aide du lemme de Silberger, à partir du caractère $\chi^\Omega$ de $\pi^\Omega$ .
D'après le corollaire 2, on a

$$(W_{23}^\Omega f)(a) = q^{-3} \sum_{a' \in U_2\backslash H_2} \sum_{u \in U} \underline{e}(u)\overline{\chi}^\Omega(h(a^{-1},0)uh(a',0)w_{23}^{-1})f(a')$$

(quels que soient $f \in M$ , $a \in G_2$), où l'on a plongé $G_2$ dans $H$ par le monomorfisme $a \mapsto h(a,0)$ $(a \in G_2)$ et l'on a identifié alors $U\backslash H$ à $U_2\backslash G_2$ . Or, chaque $u \in U$ s'écrit

$$u = h(1,b)h(u_2(s),0)\ ,$$

avec $b = (b_1,b_2) \in k^2$ , $u_2(s) = \begin{pmatrix} 1 & s \\ 0 & 1 \end{pmatrix}$ $(s \in k)$. Il s'ensuit que

$$(W^{\Omega}_{23}f)(a) = q^{-3} \sum_{\substack{a'\in U_2\backslash G_2 \\ b\in k^2 \\ s\in k}} e(s+b_2)\overline{\chi}^{\Omega}(h(1,a^{-1}b)h(a^{-1}u_2(s)a',0)w_{23}^{-1})f(a')$$

c'est-à-dire

$$(W^{\Omega}_{23}f)(a) = q^{-3} \sum_{\substack{a'\in U_2\backslash G_2 \\ b'\in k^2 \\ s\in k}} e(s+(ab')_2)\overline{\chi}^{\Omega}(h(a^{-1}u_2(s)a',b')w_{23}^{-1})f(a') \ , \tag{16}$$

quels que soient $a \in G_2$ et $f \in M$ .

4.3.3. <u>Calcul explicite des</u> $W^{\Omega}_{23}$

Rappelons que le support du caractère $\chi^{\Omega}$ est la réunion du centre $Z(G)$ de $G$ , des classes de conjugaison unipotentes

$$C_t = C\begin{pmatrix} t & 0 & 0 \\ 1 & t & 0 \\ 0 & 1 & t \end{pmatrix} \qquad (t \in k^{\times}) \ , \tag{17}$$

$$C'_t = C\begin{pmatrix} t & 0 & 0 \\ 1 & t & 0 \\ 0 & 0 & t \end{pmatrix} \qquad (t \in k^{\times}) \ , \tag{18}$$

et des classes semi-simples

$$C_z = C\begin{pmatrix} z & 0 & 0 \\ 0 & z^q & 0 \\ 0 & 0 & z^{q^2} \end{pmatrix} \qquad (z \in \mathbb{K}^{\times} \quad k^{\times}) \ , \tag{19}$$

où, bien entendu on note $C(g)$ la classe de conjugaison dans $G = GL(3,k)$ de $g \in GL(3,\mathbb{K})$ . De plus, si l'on pose

$$\mathbb{E}(z) = \mathbb{T}r(zz^q) \qquad (z \in \mathbb{K})$$

et

$$\phi_{\mathbb{K}}(z) = (\mathbb{T}r(z),\mathbb{E}(z),\mathbb{N}(z)) \qquad (z \in \mathbb{K}) \ ,$$

on a, sur les classes principales $C_z \quad (z \in \mathbb{K}^{\times})$ ,

$$\chi^{\Omega}(C_z) = \sum_{\substack{z'\in\mathbb{K} \\ \phi_{\mathbb{K}}(z')=\phi_{\mathbb{K}}(z)}} \Omega(z') \ , \tag{20}$$

et d'autre part

$$\chi^{\Omega}(C'_t) = -(q-1)\Omega(t) \qquad (t \in k^{\times}) . \tag{21}$$

Nous avons donc à décider quand un élément de la forme $h(c,b')w_{23}^{-1}$ $(c \in G_2 , b' \in k^2)$ appartient aux différentes classes $C_z (z \in \mathbb{K}^{\times})$ , $C'_t$ $(t \in k^{\times})$ .

PROPOSITION 2. <u>Avec les notations ci-dessus, on a</u>

i) <u>Si</u> $c_{21} \neq 0$ , <u>alors</u> $h(c,b')w_{23}^{-1}$ <u>ne peut appartenir qu'aux classes principales</u> $C_z$ $(z \in \mathbb{K}^{\times})$ ; <u>pour qu'il appartienne à la classe</u> $C_z$ $(z \in \mathbb{K}^{\times})$ , <u>il faut et il suffit que</u>

$$\det(c) = \mathbb{N}(z) \tag{22}$$

<u>et</u>

$$\begin{pmatrix} 0 & 1 \\ -c_{21} & c_{11} \end{pmatrix} \begin{pmatrix} b'_1 \\ b'_2 \end{pmatrix} = \begin{pmatrix} \mathbb{T}r(z) - c_{11} \\ \mathbb{E}(z) - c_{22} \end{pmatrix} ; \tag{23}$$

ii) <u>Si</u> $c_{21} = 0$ , <u>alors</u> $h(c,b')w_{23}^{-1}$ <u>ne peut appartenir qu'aux classes unipotentes; pour qu'il appartienne à</u> $C_t$ <u>ou</u> $C'_t$ $(t \in k^{\times})$ , <u>il faut et il suffit que</u>

$$\begin{cases} \det(c) = t^3 , \\ c_{11} = t , \quad c_{22} = t^2 , \\ b'_2 = 2t . \end{cases} \tag{24}$$

<u>Dans ce cas,</u> $h(c,b')w_{23}^{-1}$ <u>appartiendra à</u> $C_t$ (<u>resp.</u> $C'_t$) <u>quand</u> $b'_1 t - c_{12} \neq 0$ (<u>resp.</u> $b'_1 t - c_{12} = 0$) .

<u>Démonstration</u>: Dans le cas $c_{21} \neq 0$ , on voit aussitôt que $(h(c,b')w_{23}^{-1} - t\,\mathrm{Id})^2$ n'est jamais nul, puisque son coefficient (3,1) est justement $c_{21}$ . Ceci montre la première assertion de i). De plus, si l'on pose

$$\phi_{M(3,k)}(g) = (\mathrm{tr}(g), \mathrm{tr}(g \wedge g), \det(g)) \qquad (g \in M(3,k)) ,$$

on a

$$h(c,b')w_{23}^{-1} \in C_z \Longleftrightarrow \phi_{M(3,k)}(h(c,b')w_{23}^{-1}) = \phi_{\mathbb{K}}(z) ,$$

pour $c \in G_2$ , $b' \in k^2$ , $z \in \mathbb{K}^\times$ . Or, l'équation du membre de droite s'avère aussitôt équivalente à la conjonction de (22) et (23). La vérification de ii) est analogue et immédiate, en remarquant que la condition $b_1't - c_{12} = 0$ , en présence de (24), équivaut à $(h(c,b')w_{23}^{-1} - t\,\mathrm{Id})^2 = 0$ .

Pour calculer (16) de manière commode, nous choisissons des représentants $\underline{a}(x,t)$ $(x \in k_*^2 = k^2 - \{0\}$ , $t \in k^\times)$ pour $U_2 \backslash G_2$ comme suit,

$$\underline{a}(x,t) = \begin{pmatrix} 0 & -tx_1^{-1} \\ x_1 & x_2 \end{pmatrix} \qquad \text{si} \quad x_1 \neq 0 , \tag{25}$$

$$\underline{a}(x,t) = \begin{pmatrix} tx_2^{-1} & 0 \\ 0 & x_2 \end{pmatrix} \qquad \text{si} \quad x_1 = 0 . \tag{26}$$

Alors la classe latérale droite $U_2\underline{a}(x,t)$ dans $G_2$ consiste des matrices de déterminant $t$ et de seconde ligne $x$ .

La formule (16) s'écrit maintenent

$$(W_{23}^{\Omega}f)(\underline{a}(x,t)) = \tag{27}$$

$$= q^3 \sum_{\substack{y \in k_*^2 \\ r \in k^\times \\ b' \in k^2 \\ s \in k}} e(s+x\cdot b')\overline{\chi}^{\Omega}(h(c(x,t;s;y,r),b')w_{23}^{-1})f(\underline{a}(y,r)) ,$$

où l'on a posé $c(x,t;s;y,r) = \underline{a}(x,t)^{-1}u_2(s)\underline{a}(y,r)$ pour $x,y \in k_*^2, r,t \in k^\times$, $s \in k$ .

Le cas $x_1 \neq 0$ .

Décomposons la sommation du membre de droite de (27) sous la forme

$$S^{**} + S^{*o} + S^{o}$$

où $S^{**}$ (resp. $S^{*o}$ ; resp. $S^{o}$) désigne la sous-somme portant sur les

$y \in k_*^2$ avec $y_1 \neq 0$ , $r \in k^\times$ , $b' \in k^2$ et $s \in k^\times$ (resp. les $y \in k_*^2$ avec $y_1 \neq 0$ , $r \in k^\times$ , $b' \in k^2$ et $s = 0$ ; resp. $y = (0,y_2)$ , $y_2 \in k^\times$ , $r \in k^\times$ , $b' \in k^2$ et $s \in k$ ).

Pour calculer $S^{**}$ , on remarque que, dans ce cas,

$$c(x,t;s;y,r)_{11} = y_1(x_2 st^{-1} + x_1^{-1}) ,$$

$$c(x,t;s;y,r)_{21} = -st^{-1}x_1y_1 ,$$

$$c(x,t;s;y,r)_{22} = x_1(rt^{-1}y_1^{-1} - st^{-1}y_2) ,$$

quels que soient $x,y \in k_*^2$ avec $x_1,y_1 \neq 0$ , $r,t \in k^\times$ , $s \in k$ . D'après la proposition 2,i), on trouve alors aussitôt que pour

$$h(c(x,t;s;y,r),b')w_{23}^{-1} \in C_z \qquad (z \in \mathbb{K}^\times) ,$$

on a

$$x\cdot b' = -s^{-1}tx_1^{-2}y_1^{-2}(x_1^3\,\mathbb{N}(z) - x_1^2y_1\,\mathbb{E}(z) + x_1y_1^2\,\mathbb{T}r(z) - y_1^3 - st^{-1}y_1^3x_1x_2 - st^{-1}x_1^3y_1y_2)$$

$$= -s^{-1}tx_1^{-2}y_1^{-2}[\mathbb{N}(x_1z - y_1)] + [y_1x_1^{-1}x_2 + x_1y_1^{-1}y_2] .$$

On en tire, compte tenu de (20), que

$$S^{**} = q^{-3} \sum_{\substack{y\in k^2 \\ y_1\neq 0 \\ s\in k^\times \\ z\in\mathbb{K}^\times}} e(s - s^{-1}tx_1^{-2}y_1^{-2}\mathbb{N}(x_1z - y_1) + y_1x^{-1}x_2 + x_1y_1^{-1}y_2)\cdot \\ \cdot\overline{\Omega}(z)f(\underline{a}(y,t\mathbb{N}(z))). \qquad (28)$$

Pour calculer $S^{*o}$ , notons que dans ce cas $(x,y \in k_*^2$ , $x_1,y_1 \neq 0$ , $s = 0$ , $r \in k^\times)$ , on a

$$c(x,t;0;y,r) = \begin{pmatrix} x_1^{-1}y_1 & -rt^{-1}x_2y_1^{-1} + x_1^{-1}y_2 \\ 0 & rt^{-1}x_1y_1^{-1} \end{pmatrix} .$$

D'apres la prop. 2,ii), on trouve que pour $h(c(x,t:0:y,r),b')w_{23}^{-1} \in C'_{t'}$ , on a $t' = y_1x_1^{-1}$ et

$$x\cdot b' = t'x_2 + t'^{-1}y_2 = y_1x_1^{-1}x_2 + x_1y_1^{-1}y_2 \quad .$$

D'autre part, pour $h(c(x,t;0;y,r),b')w_{23}^{-1} \in C_{t'}$ , on a aussi $t' = y_1x_1^{-1}$ , mais

$$x\cdot b' = x_1b_1' + 2x_2t' = x_1b' + 2y_1x_1^{-1}x_2 \ ,$$

où $b_1' \neq -t'x_1^{-1}x_2 + t'^{-1}x_1^{-1}y_2$ . Il s'ensuit, compte tenu de (20) et (21) ,

$$S^{*o} = -(q-1) \sum_{\substack{y\in k^2 \\ y_1\neq 0}} e(y_1x_1^{-1}x_2 + x_1y_1^{-1}y_2)\overline{\hbar}(y_1x_1^{-1})f(\underline{a}(y,ty_1^3x_1^{-3})) +$$

$$+ q^{-3} \sum_{\substack{y\in k^2 \\ y_1\neq 0 \\ b_1'\in k \\ x_1b_1'\neq y_1x_1^{-1}x_2+x_1y_1^{-1}y_2}} e(x_1b_1' + 2y_1x_1^{-1}x_2)\overline{\hbar}(y_1x_1^{-1})f(\underline{a}(y,ty_1^3x_1^{-3})) \ ,$$

c'est-à-dire

$$S^{*o} = -q^{-2} \sum_{\substack{y\in k^2 \\ y_1\neq 0}} e(x_1y_1^{-1}y_2 + y_1x_1^{-1}x_2)\overline{\hbar}(y_1x_1^{-1})f(\underline{a}(y,ty_1^3x_1^{-3})) \ . \qquad (29)$$

Pour calculer $S^o$ , remarquons que dans ce cas $(x,y \in k_*^2$ , $x_1 \neq 0$ , $y_1 = 0$ , $s \in k$ , $r,t \in k^\times)$ , on a

$$c(x,t;s;y,r) = t^{-1}\begin{pmatrix} rx_2y_2^{-1} & sx_2y_2 + ty_2x_1^{-1} \\ -rx_1y_2^{-1} & -sx_1y_2 \end{pmatrix} .$$

Il découle alors de la prop. 2,i) que, pour $h(c(x,t;s;y,r),b')w_{23}^{-1} \in C_z \quad (z \in \mathbb{K}^\times)$ , on a

$$x\cdot b' = y_2\mathbb{N}(z)^{-1}\mathbb{E}(z) = st^{-1}\mathbb{N}(z)^{-1}x_1y_2^2 \ .$$

D'après (20) et la prop. 2,i), on trouve

$$S^{o} = q^{-3} \sum_{\substack{y\in k_*^2 \\ y_1=0 \\ z\in K^\times \\ t\mathbb{N}(z)=-x_1y_2^2}} e(s + y_2\mathbb{N}(z)^{-1}\mathbb{E}(z) + st^{-1}\mathbb{N}(z)x_1y_2^2)\overline{\Omega}(z)f(\underline{a}(y,t\mathbb{N}(z))) ,$$

c'est-à-dire

$$S^{o} = q^{-2} \sum_{\substack{y\in k_*^2 \\ y_1=0 \\ z\in K^\times \\ \mathbb{N}(z)y_2^{-2}=-t^{-1}x_1}} e(-tx_1^{-1}y_2^{-1}\mathbb{E}(z))\overline{\Omega}(z)f(\underline{a}(y,t\mathbb{N}(z))) . \tag{30}$$

En introduisant la fonction de Bessel $J_o$ sur $k$ , définie par

$$J_o(t) = \sum_{\substack{r,s\in k \\ rs=t}} e(r - s) \qquad (t \in k ) , \tag{31}$$

$$J_o(0) = -(q + 1) , \tag{32}$$

on obtient alors, compte tenu de (28), (29) et (30) , la

PROPOSITION 3. _Pour_ $x \in k_*^2$ _avec_ $x_1 \neq 0$ , $t \in k^\times$ _et_ $f \in M$ , _on a_

$$(W_{23}^{\Omega}f)(\underline{a}(x,t)) =$$

$$= q^{-3} \sum_{\substack{y\in k^2 \\ z\in K^\times}} J_o(tx_1^{-2}y_1^{-2}\mathbb{N}(x_1z - y_1))e(y_1x_1^{-1}x_2 + x_1y_1^{-1}y_2)\overline{\Omega}(z)f(\underline{a}(y,t\mathbb{N}(z))) +$$

$$+ q^{-2} \sum_{\substack{z\in K^\times \\ y_2\in k^\times \\ \mathbb{N}(z)y_2^{-2}=-x_1t^{-1}}} e(-tx_1y_2^{-1}\mathbb{E}(z))\overline{\Omega}(z)f(\underline{a}((0,y_2),t\mathbb{N}(z))) .$$

_Le cas_ $x_1 = 0$ .

De manière analogue au cas précédent, on peut décomposer la sommation du membre de droite dans (27) sous la forme $S^* + S^o$ , où, maintenant, $S^*$ (resp. $S^o$) désigne la sous-somme portant sur les $y \in k_*^2$ avec $y_1 \neq 0$ , $r \in k^\times$ , $b' \in k^2$ et $s \in k$ (resp. les $y \in k_*^2$ avec $y_1 = 0$ , $r \in k^\times$ , $b' \in k^2$ et $s \in k$) .

Pour calculer $S^*$ , notons que dans ce cas

$$c(x,t;s;y,r)_{11} = st^{-1}x_2y_1 ,$$

$$c(x,t;s;y,r)_{21} = y_1x_2^{-1} .$$

D'après la prop. 2,i) et (20), on trouve aussitôt

$$S^* = q^{-3} \sum_{\substack{y\in k_*^2 \\ y_1\neq 0 \\ z\in K^\times \\ s\in k}} e(s + x_2 \mathbb{T}r(z) - x_2^2 y_1 st^{-1}) \bar{\Omega}(z) f(\underline{a}(y, t\mathbb{N}(z))) ,$$

c'est-à-dire

$$S^* = q^{-2} \sum_{\substack{y_2\in k^\times \\ z\in K^\times}} e(x_2 \mathbb{T}r(z)) \bar{\Omega}(z) f(\underline{a}((tx_2^{-2}, y_2), t\mathbb{N}(z))) . \qquad (33)$$

Pour calculer $S^o$ , remarquons que dans ce cas

$$c(x,t;s;y,r) = \begin{pmatrix} rt^{-1}x_2y_2^{-1} & st^{-1}x_2 \\ 0 & y_2x_2^{-1} \end{pmatrix}$$

et que, d'après la prop. 1,ii), $h(c(x,t;s;y,r),b')w_{23}^{-1}$ appartient à $C_{t_o} \cup C'_{t_o}$ si et seulement si

$$r = tt_o^3 , \quad y_2 = t_o^2 x_2 \quad \text{et} \quad b'_2 = 2t_o .$$

Il s'ensuit que

$$S^o = q^{-3} \sum_{\substack{t_o\in k^\times \\ s\in k \\ b'_1\in k}} e(s + 2t_o x_2) \bar{\Omega}(t_o) f(\underline{a}((0, t_o^2 x_2), tt_o^3)) = 0 .$$

On a ainsi obtenu la

PROPOSITION 4. Pour $x \in k_*^2$ avec $x_1 = 0$, $t \in k^\times$ et $f \in M$, on a

$$(W_{23}^{\Omega}f)(a\_(x,t)) = q^{-2} \sum_{\substack{y_2 \in k^\times \\ z \in \mathbb{K}}} e(x_2 \mathbb{T}r(z))\overline{\Omega}(z)f(\underline{a}((tx_2^{-2},y_2), t\mathbb{N}(z))) .$$

Ceci complète le calcul explicite des modèles de Kirillov de la série discrète de $G$ à l'aide des caractères correspondants. Cette méthode de construction a été employeé pour la première fois, pour le groupe $G$, dans [HP], où l'on trouve une présentation équivalente à celle que nous donnons ici, mais beaucoup moins adaptée à la construction de "représentations de Weil généralisées" pour $G$. Un calcul explicite de modèles pour la série discrète de $G$ à l'aide des fonctions de Bessel $J^{\Omega}$ a été donné auparavant dans [G]; ces modèles semblent néanmois d'un maniement moins aisé que ceux de [HP] ou ceux que nous donnons ici.

### 4.4. La "représentation de Weil généralisée" de $G$ associée à l'extension cubique $\mathbb{K}$

#### 4.4.1. Plongement des modèles de Kirillov dans une représentation commune

Nous montrons tout d'abord comment plonger les différents modèles de Kirillov $(M,\pi^{\Omega})$, que nous venons de construire, dans une grosse représentation $(V',\rho')$ de $G$, dont l'action $\rho'$ est plus simple que chaque $\pi^{\Omega}$.

Rappelons que l'espace $M$ est formé des fonctions $f$ de $G_2$ dans $\mathbb{C}$, telles que (cf. 2.2.)

$$f(u_2(s)c) = e(s)f(c) \qquad (s \in k\ ,\ c \in G_2) .$$

Nous associons à chaque $f \in M$ une fonction, notée $\tilde{f}$, de $\mathbb{K}^\times \times G_2$ dans $\mathbb{C}$, définie par

$$\tilde{f}(z,a) = \overline{\Omega}(z)f(d(\mathbb{N}(z),1)a) \qquad (z \in \mathbb{K}^\times\ ,\ a \in G_2) , \tag{34}$$

où nous posons

$$d(r,s) = \begin{pmatrix} r & 0 \\ 0 & s \end{pmatrix} \qquad (r,s \in k^\times) .$$

On a alors

$$f(a) = \widetilde{f}(1,a) \qquad (a \in G_2) ,$$

et la fonction $\widetilde{f}$ vérifie les équations fonctionnelles

$$\widetilde{f}(z'z, d(\mathbb{N}(z')^{-1},1)a) = \overline{\Omega}(z')\widetilde{f}(z,a) \qquad (z,z' \in \mathbb{K}^\times , \quad a \in G_2) \tag{35}$$

et

$$\widetilde{f}(z,u_2(s)a) = e(s\mathbb{N}(z))\widetilde{f}(z,a) \qquad (z \in \mathbb{K}^\times , \ s \in k , \ a \in G_2) . \tag{36}$$

Le théorème suivant découle alors aussitôt de ce qui precède.

THEOREME 1. Avec les notations ci-dessus, la correspondance $f \mapsto \widetilde{f}$ établit un isomorphisme entre notre modèle de Kirillov $(M,\pi^\Omega)$ et la représentation $(V'_\Omega,\rho')$, dont l'espace $V'_\Omega$ est formé des fonctions $\widetilde{f}$ de $\mathbb{K}^\times \times G_2$ dans $\mathbb{C}$ satisfaisant (35) et (36), et dont l'action $\rho'$ est donnée par les formules

$$(\rho'_r \widetilde{f})(z,a) = \widetilde{f}(r^{-1}z, d(r^3,1)a) \qquad (r \in Z(G) \simeq k^\times) , \tag{37}$$

$$(\rho'_{h(1,b)}\widetilde{f})(z,a) = e((ab)_2)\widetilde{f}(z,a) \qquad (b \in k^2) , \tag{38}$$

$$(\rho'_{h(a',0)}\widetilde{f})(z,a) = \widetilde{f}(aa') \qquad (a' \in G_2) , \tag{39}$$

$$(\rho'_{w_{23}} \widetilde{f})(z,\underline{a}(x,t)) = \tag{40}$$

$$= q^{-3} \sum_{\substack{z' \in \mathbb{K}^\times \\ y \in k_*^2 \\ y_1 \neq 0}} J_o(x_1^{-2}y_1^{-2}t\mathbb{N}(x_1 z' - y_1 z))e(y_1 x_1^{-1} x_2 + x_1 y_1^{-1} y_2)\widetilde{f}(z',\underline{a}(y,t)) +$$

$$+ q^{-2} \sum_{\substack{z' \in \mathbb{K}^\times \\ y \in k_*^2, y_1 = 0 \\ \mathbb{N}(z') y_2^{-2} = -x_1 t^{-1} \mathbb{N}(z)}} e(-t x_1^{-1} y_2^{-1} \mathbb{E}(z' \bar{z}' \bar{\bar{z}})) \tilde{f}(z', \underline{a}(y,t))$$

si $x_1 \neq 0$,

$$(\rho'_{w_{23}} \tilde{f})(z, \underline{a}(x,t)) = q^{-2} \sum_{\substack{z' \in \mathbb{K}^\times \\ y_2 \in k^\times}} e(x_2 \mathbb{T}r(z' z^{-1})) \tilde{f}(z, \underline{a}((t\mathbb{N}(z) x_2^{-2}, y_2), t)) \tag{41}$$

si $x_1 = 0$, quels que soient $\tilde{f} \in V_\Omega$, $z \in \mathbb{K}^\times$, $x \in k_*^2$, $t \in k^\times$.

4.4.2. La "représentation de Weil généralisée" $(V', \rho')$

Le théorème précédent fait apparaître les modèles de Kirillov $(M, \pi^\Omega)$ de la série discrète de $G$ comme des sous-représentations de la représentation $(V', \rho')$ de $G$ dans l'espace $V'$ formé des fonctions $f'$ de $\mathbb{K}^\times \times G_2$ dans $\mathbb{C}$ telles que

$$f'(z, u_2(s)a) = e(s\mathbb{N}(z)) f(z,a) \qquad (z \in \mathbb{K}^\times,\ s \in k^\times,\ a \in G_2) \tag{42}$$

et

$$\sum_{u \in \mathbb{U}} f'(uz, a) = 0, \tag{43}$$

où $\mathbb{U}$ désigne le noyau de la norme $\mathbb{N}$, dont l'action $\rho'$ est donnée par les formules (37) à (41), pour $f'$ à la place de $\tilde{f}$.

C'est donc cette représentation $(V', \rho')$ que nous appelerons (provisoirement) "représentation de Weil généralisée" de $G$ associée à l'extension cubique $\mathbb{K}$ de $k$. Elle se décompose suivant l'action naturelle de $\mathbb{K}^\times$, définie par

$$(z \cdot f')(z', a) = f'(zz', d(\mathbb{N}(z)^{-1}, 1)a) \tag{44}$$

pour $f' \in V'$, $z, z' \in \mathbb{K}^\times$, $a \in G_2$. Plus précisement, ses composantes irréductibles, à savoir les $V'_\Omega$ $(\Omega \in (\mathbb{K}^\times)^\wedge - (k^\times)^\wedge)$ , sont données par (35); les seuls cas d'isomorphie non triviales,

$$V_\Omega \simeq V_{\Omega^q} \simeq V_{\Omega^{q^2}}$$

proviennent del'action naturelle du groupe de Galois de $\mathbb{K}$ sur $k$ dans $(V',\rho')$ . En fait, $(V',\rho')$ se décompose sans multiplicités suivant le groupe de similitudes $GO(\mathbb{N})$ de la forme cubique $\mathbb{N}$ , isomorphe au produit semidirect de $\mathbb{K}^\times$ et du groupe de Galois de $\mathbb{K}$ sur $k$ .

4.4.3. Questions ouvertes

i) Comme dans le cas quadratique, on peut envisager de construire la représentation $(V',\rho')$ en donnant l'action d'un sistème de générateurs convenable de $G$ , et en vérifiant ensuite que les relations entre ces générateurs sont respectées.

ii) Ensuite, il reste à construire des modèles analogues à $(V',\rho')$ , pour les autres séries de représentations de $G$ .

iii) On voudrait étendre ces constructions à $GL(n,k)$ , aussi dans le cas d'un corps local.

iv) Enfin, on conjecture qu'on n'a ici que l'ébauche d'une nouvelle méthode de construction des représentations des groupes classiques.

Bibliographie

[ A-S ] ATIYAH, M.F., et SCHMID, W., A geometric construction of the discrete series for semi-simple Lie groups, Invent. Math. 42 (1977), 1-62.

[ D-L ] DELIGNE, P. et LUSZTIG, G., Representations of reductive groups over finite fields, Ann. of Math. 103 (1976), 103-161.

[G ] GELFAND, S.I., Representations of the Full General Group over a Finite Field, Mat. Sbornik, 83 (125), 1970, n° 1; Math. USSR Sbornik, 12, 1970, n° 1.

[ H-C 1] HARISH-CHANDRA, Harmonic Analysis on semisimple Lie Groups, Bull. Amer. Math. Soc. 78 (1970), 529-551.

[H-C 2] HARISH-CHANDRA, Harmonic Analysis on reductive p-adic groups, in Harmonic Analysis on Homogeneous Spaces, Proc. Symp. Pure Maths. Vol. 26 (1973), A.M.S., Providence, 1973, p. 167-192.

[H-C 3] HARISH-CHANDRA, Eisenstein series over finite fields, Functional Analysis and related fields, Springer-Verlag, Berlin & New York, 1970, p. 76-88 .

[HP] HELVERSEN-PASOTTO, A., Darstellungen von $GL(3,\mathbb{F}_q)$ , à paraître aux Math. Ann.

[L-V] LION, G. et VERGNE, M., The Weil representation, Maslov index and Theta series, Birkhäuser, Boston, Basel, Stuttgart, 1980.

[S 1] SCHMID, W., $L^2$ cohomology and the discrete series, Ann. of Math., 103 (1976), p. 375-394.

[S 2] SCHMID, W., Representations of Semisimple Lie groups, Proc. Intern. Congress of Mathematicians, Helsinki, 1978, p. 195-207.

[SA 1] SOTO ANDRADE, J., Représentations de certains groupes symplectiques, Mémoire 55-56, Soc. Math. de France, 1978.

[SA 2] SOTO ANDRADE, J., Représentations de Weil généralisées, preprint, Centre de Mathématiques, École Polytechnique, Palaiseau, 1982.

[W] WEIL, A., Sur certains groupes d'opérateurs unitaires, Acta Math., 111 (1967), p. 143-211.

# 9
# Some New Families of Chromatically Unique Graphs

REINALDO E. GIUDICI / Departamento de Matemática y Ciencia de la Computación, Universidad Simon Bolivar, Caracas, Venezuela

## 1. INTRODUCTION

The graphs considered in this paper are always finite, undirected, loopless and without multiple lines, i.e., graphs as defined in [5]; in general we shall follow the terminology used there.

A *coloring* of a graph is an assignment of one of a specified set $\{1, 2, 3, \ldots, \lambda\}$ of colors to each point of the graph so that *no two adjacent points have the same color*. It is however not required that all the $\lambda$ colors at our disposal be really used. It is a well known fact (see [5, Corollary 12.32(a)] or [8]) that the number of different colorings of a graph G in which at most $\lambda$ colors are used, is a polynomial in $\lambda$, called the *chromatic polynomial* (or *chromial*) of G. In this paper it will be denoted by $P(G,\lambda)$; its principal properties used in this paper will be summarized below in Section 3.

Nonisomorphic graphs can have the same chromatic polynomial; for instance, all the trees $T_p$ with the same number p of points have the same chromatic polynomial

$$P(T_p,\lambda) = \lambda(\lambda - 1)^{p-1} = \lambda(\lambda - 1)^q \quad (1.1)$$

see, e.g., [8, Theorem 5], where $q = p - 1$ represents the number of edges.

There are however graphs which are completely characterized (up to isomorphism) by their chromatic polynomial, i.e., graphs G such that if for any graph H

$$P(H,\lambda) = P(G,\lambda) \tag{1.2}$$

holds, then H is isomorphic to G. Such graphs are called *chromatically unique*. Several families of such graphs are known; those we have found published are summarized below in Section 2.

No general method for finding such families nor for proving chromatic uniqueness seems to be known. Here we study the possibilities of deriving new chromatically unique graphs from the graphs with the same property of uniqueness, for example, by adding isolated points (Section 4) or a line ending in a pendant point (Section 5). In the same section (see Theorem 5.1) a new 2-parameter family of chromatically unique graphs is also described; in their complement any graph can be characterized as the union of a star and certain number of isolated points.

## 2. KNOWN FAMILIES OF CHROMATICALLY UNIQUE GRAPHS

1. The complete graphs $K_p$ having the chromatic polynomial

$$P(K_p,\lambda) = (\lambda)_p \tag{2.1}$$

where $(\lambda)_p$ stands for the "descending factorial"

$$(\lambda)_p = \lambda(\lambda - 1)(\lambda - 2)\cdots(\lambda - p + 1) \tag{2.2}$$

2. The complement $\bar{K}_p = pK_1$ with

$$P(\bar{K}_p,\lambda) = \lambda^p \tag{2.3}$$

3. The cycles $C_p$ with

$$P(C_p,\lambda) = (\lambda - 1)\{(\lambda - 1)^{p-1} + (-1)^p\} \tag{2.4}$$

see [1].

4. Cycles $C_p$ with a chord, called $\theta$-graphs in [1].
5. The generalized $\theta$-graphs $\theta_{d,e,f}$ described in [6,7].
6. Chao and Whitehead, Jr. showed in [2] that the wheel $W_6$ is not chromatically unique (nothing seems to be known about wheels with more than six points), but that chromatically unique graphs can be obtained from wheels by deleting certain "spokes." For instance, such a family $X_p$ ($p \geq 5$) results if we maintain in a wheel only *three* consecutive spokes; for

$P(X_p,\lambda)$ the following formula is given in [2]:

$$P(X_p,\lambda) = (\lambda - 2)\{(\lambda - 1)^{p-1} - (\lambda - 1)^{p-2} + (-1)^{p-1} \cdot 2(\lambda - 1)\} \quad (2.5)$$

7. In the same paper [2] also the graphs $Z_p$ ($p \geq 6$) resulting in an analogous fashion by maintaining in a wheel *four* consecutive spokes are shown to be chromatically unique; for their chromatical polynomial, Loerinc [6, p. 102] gives the following explicit formula:

$$P(Z_p,\lambda) = (\lambda - 1)(\lambda - 2)\{(\lambda - 2)^2(\lambda - 1)^{p-4} + (-1)^{p-4}(\lambda - 4)\} \quad (2.6)$$

8. Graphs resulting from a $K_p$ by deleting any number k (where k = 1, 2, ..., [p/2]) of mutually disjoint lines; these graphs might be described as $\overline{kK_2 \cup (p - 2k)K_1}$. See [6, Theorem 5.3.2, attributed to E. G. Whitehead, Jr.], where chromatic uniqueness is proved. The chromatic polynomial (not given there) can be easily found by using "Property VII" in the next section; in the notation introduced in (2.2) it can be written as

$$P(\overline{kK_2 \cup (p - 2k)K_1}, \lambda) = \sum_{i=0}^{k} \binom{k}{i} (\lambda)_{p-i} \quad (2.7)$$

or also as

$$P(\overline{kK_2 \cup (p - 2k)K_1}, \lambda) = (\lambda)_{p-k} \sum_{i=0}^{k} \binom{k}{i} (\lambda - p + k)_{k-i} \quad (2.8)$$

since it is easily checked that for $0 \leq i \leq k$ the following identity holds:

$$(\lambda)_{p-k}(\lambda - p + k)_{k-i} = (\lambda)_{p-i} \quad (2.9)$$

## 3. SUMMARY OF SOME PROPERTIES OF CHROMATIC POLYNOMIALS

I. For any graph G with p points and q lines the chromatic polynomial is

$$P(G,\lambda) = \lambda^p - q\lambda^{p-1} + \cdots \quad (3.1)$$

II. If $P(G,\lambda)$ is divisible by $\lambda^k$, but not by $\lambda^{k+1}$, then G has exactly k connected components.

III. G has at least one line iff $P(G,\lambda)$ is divisible by $\lambda(\lambda - 1)$.

COROLLARY. The only chromatic polynomial not divisible by $\lambda - 1$ is the power $\lambda^p$; see (2.3) above.

IV. If

$$G = G_1 \cup G_2 \cup \cdots \cup G_k \tag{3.2}$$

where the symbol $\cup$ stands (as always in this paper) for disjoint union, then

$$P(G,\lambda) = \prod_{i=1}^{k} P(G_i,\lambda) \tag{3.3}$$

V. (Whitney [9]): If A and B are a pair of nonadjacent vertices in G, let G' be the graph obtained from G by the addition of a line AB, and G'' be obtained by contracting A and B to one vertex; then

$$P(G,\lambda) - P(G',\lambda) = P(G'',\lambda) \tag{3.4}$$

VI. (Zykov [10]): If two graphs X and Y "overlap" in a complete graph $K_n$, then the chromatic polynomial of the graph G formed by X and Y together is

$$P(G,\lambda) = \frac{P(X,\lambda) \cdot P(Y,\lambda)}{(\lambda)_n} \tag{3.5}$$

where (as always in this paper)

$$(\lambda)_n = \lambda(\lambda - 1)(\lambda - 2)\cdots(\lambda - n + 1) \tag{3.6}$$

The meaning of "overlapping" is explained in [8, p. 59] as follows: the sets V(X) and V(Y) are not disjoint, but have n points in common, and every pair of these n points is adjacent both in X and Y.

VII. (Frucht [3]): For any graph H with p points and any k = 1, 2, ..., p define $b_k(H)$ as the *number of spanning* subgraphs of H which have k connected components each of which is a *complete* graph (that is, the subgraphs to be considered are of the form

$$K_{\alpha_1} \cup K_{\alpha_2} \cup \cdots \cup K_{\alpha_k} \qquad \text{where } \alpha_1 + \alpha_2 + \cdots + \alpha_k = p) \tag{3.7}$$

Then for any graph G with p points we have

$$P(G,\lambda) = \sum_{k=1}^{p} b_k(\bar{G})(\lambda)_k \tag{3.8}$$

where $\bar{G}$ is (as usual) the complement of G; for $(\lambda)_k$ see (3.6) above.

In particular we have for k = p - 1 that

$$b_{p-1}(\bar{G}) = \bar{q} \tag{3.9}$$

(the number of lines in $\bar{G}$), since in this case (3.7) reduces to $(p - 2)K_1 \cup K_2$.

## 4. ADDING ISOLATED POINTS TO A CHROMATICALLY UNIQUE GRAPH

The union

$$G = H \cup \bar{K}_n = H \cup nK_1 \tag{4.1}$$

of a chromatically unique graph H and $\underline{n}$ isolated points (where $\underline{n}$ is any point integer) is by no means always again chromatically unique. As an instance already for n = 1 take as H the graph of Fig. 1; we shall call this graph $K_3^+$ because it is obtained from a triangle $K_3$ by adding a fourth point which is then joined by a line to a point of the triangle (to which one is indifferent since $K_3$ is point-symmetric). A pictorial description of the graph $K_3^+$ might be as follows: we take a copy of $K_3$ as if it were a picture and hang it on the wall of our room with the aid of a nail and thread (see Fig. 1). (It goes without saying that the same operation can be effectuated without ambiguity for any connected *point-symmetric* graph; this possibility will be considered below in Section 5).

FIGURE 1. The graph $K_3^+$.

That $K_3^+$ is chromatically unique follows from Theorem 5.3 below or from [4], where this graph appears as the only existing one with the chromatic polynomial

$$P(K_3^+,\lambda) = \lambda(\lambda - 1)^2(\lambda - 2) \tag{4.2}$$

The graph $K_3^+ \cup K_1$ with the chromatic polynomial

$$P(K_3^+ \cup K_1, \lambda) = \lambda^2(\lambda - 1)^2(\lambda - 2) \tag{4.3}$$

however is not chromatically unique; in fact, the graph $K_3 \cup K_2$ has the same chromatic polynomial (4.3) as can be seen from [4, p = 5, q = 4, #5] or by applying Property IV to (2.1).

It is however possible to assure chromatic uniqueness of the graph (4.1) by adding the condition that $P(H,\lambda)$ not be divisible by $(\lambda - 1)^2$. In fact, we shall now prove the following theorem.

THEOREM 4.1. If a graph H is chromatically unique and its chromatic polynomial $P(H,\lambda)$ is not divisible by $(\lambda - 1)^2$, then the graphs $H \cup \bar{K}_n$ are also chromatically unique for n = 1, 2, 3, ....

PROOF. By induction:

I. The theorem is true for n = 1. In fact, it follows from (2.3) and Property IV that

$$P(H \cup K_1, \lambda) = P(H,\lambda) \tag{4.4}$$

Now let G be any graph such that

$$P(G,\lambda) = \lambda P(H,\lambda) \tag{4.5}$$

we have to show that G is isomorphic to $H \cup K_1$. Because of (2.3) the case that $P(H,\lambda)$ is a power of $\lambda$ is trivial. Hence we can suppose (using the hypothesis and Property III) that

$$P(H,\lambda) = \lambda^{\alpha}(\lambda - 1)Q(\lambda) \qquad \alpha \geq 1 \tag{4.6}$$

where $Q(\lambda)$ is neither divisible by $\lambda$ nor by $\lambda - 1$. Hence:

$$P(G,\lambda) = \lambda^{\alpha+1}(\lambda - 1)Q(\lambda) \tag{4.7}$$

It follows now from Property II that G is the union of $\alpha + 1$ connected components (where $\alpha + 1 \geq 2$), say

$$G = G_1 \cup G_2 \cup \cdots \cup G_{\alpha+1} \tag{4.8}$$

and because of Property IV we have

$$P(G,\lambda) = P(G_1,\lambda) \cdot P(G_2,\lambda) \cdot \cdots \cdot P(G_{\alpha+1},\lambda)$$

But from (4.7) we see that only *one* of the $P(G_i,\lambda)$ can be divisible by $\lambda - 1$; for the sake of simplicity, we suppose i = 1. Then it follows from the corollary to Property III that

$$\begin{cases} P(G_1,\lambda) = \lambda(\lambda - 1)Q(\lambda) \\ P(G_2,\lambda) = P(G_3,\lambda) = \cdots = P(G_{\alpha+1},\lambda) = \lambda \end{cases} \tag{4.10}$$

Hence (4.8) becomes

$$G = G_1 \cup \bar{K}_{\alpha} = G_1 \cup \alpha K_1 \tag{4.11}$$

or also

$$G = (G_1 \cup \overline{K_{\alpha-1}}) \cup K_1 \tag{4.12}$$

It now follows from (4.5) and Property IV that

$$P(G_1 \cup \overline{K_{\alpha-1}}, \lambda) = \frac{P(G,\lambda)}{\lambda} = P(H,\lambda) \tag{4.13}$$

hence, by the uniqueness hypothesis, $G_1 \cup \overline{K_{\alpha-1}}$ is isomorphic to H, and so, because of (4.12), G is isomorphic to $H \cup K_1$. Q.E.D.

II. It follows now easily that the theorem is true for $H \cup \overline{K_{n+1}}$ if it has already been proved for $H \cup \bar{K}_n$. Indeed,

$$H \cup \bar{K}_{n+1} = (H \cup \bar{K}_n) \cup K_1 \tag{4.14}$$

and this is the case I applied to $H \cup \bar{K}_n$ (instead of H). [Indeed, it follows from Property IV that $P(H \cup \overline{K_{n+1}}, \lambda)$ is not divisible by $(\lambda - 1)^2$ if $P(H \cup \bar{K}_n, \lambda)$ is not.] ∎

Theorem 4.1 yields families $H \cup \bar{K}_n = H \cup nK_1$ of chromatically unique graphs if we use for H the graphs described in Section 2 whenever it can be shown that $P(H,\lambda)$ is not divisible by $(\lambda - 1)^2$

1. $K_m \cup \bar{K}_n$ is chromatically unique with

$$P(K_m \cup \bar{K}_n, \lambda) = \lambda^{n+1}(\lambda - 1)(\lambda - 2)\cdots(\lambda - m + 1) \tag{4.15}$$

2. $\bar{K}_m \cup \bar{K}_n = \overline{K_{m+n}}$ is trivially chromatically unique.
3. From (2.4) it can be seen that $P(C_m,\lambda)$ is not divisible by $(\lambda - 1)^2$; hence also the graphs $C_m \cup \bar{K}_n$ are chromatically unique, with

$$P(C_m \cup \bar{K}_m, \lambda) = \lambda^n(\lambda - 1)\{(\lambda - 1)^{m-1} + (-1)^m\} \quad m \geq 3,\ n \geq 1 \tag{4.16}$$

4. As can be seen from Figure 2 (drawn for the example e = 4, f = 5), the elimination of the chord AB in the graph $\theta_{e,f}$ yields a cycle $C_{e+f-2}$,

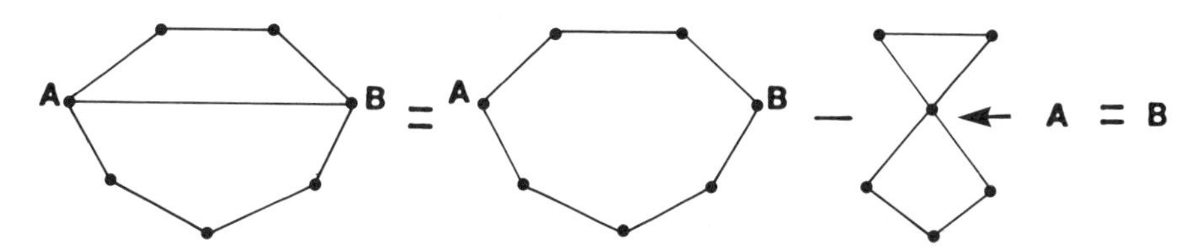


FIGURE 2. The graph $\theta_{e,f}$ for e = 4, f = 5, and the computation of its chromatic polynomial.

while the contraction of the two endpoints of this chord into a single point leads to two cycles, $C_{e-1}$ and $C_{f-1}$, "overlapping" in a $K_1$. Hence it follows from Properties V and VI that

$$P(\theta_{e,f},\lambda) = P(C_{e+f-2},\lambda) - \frac{P(C_{e-1},\lambda) \cdot P(C_{f-1},\lambda)}{\lambda} \tag{4.17}$$

where $e \geq 3$, $f \geq 3$; if an equality sign holds, the arising $C_2$ in the numerator of the fraction is to be replaced by $K_2$. At all events this fraction is a polynomial divisible by $(\lambda - 1)^2$; but $P(C_{e+f-2},\lambda)$ is divisible only by $(\lambda - 1)$ hence this holds also for the difference. This reasoning suffices to show that $\theta_{e,f}$ satisfies the hypothesis of Theorem 4.1. Hence the graphs $\theta_{e,f} \cup \bar{K}_n$ are chromatically unique.

5. In a similar fashion it can be shown that also $P(\theta_{d,e,f},\lambda)$ is not divisible by $(\lambda - 1)^2$; see [7, p. 314]. Hence also the graphs $\theta_{d,e,f} \cup \bar{K}_n$ are chromatically unique.
6. The same holds for the graphs $X_m \cup \bar{K}_n$ $(m \geq 5)$, since (2.5) shows that the polynomial $P(X_p,\lambda)$ is not divisible by $(\lambda - 1)^2$.
7. Id. for the graphs $Z_m \cup \bar{K}_n$ $(m \geq 6)$; see (2.6).
8. The graphs $\overline{kK_2 \cup (m - 2k)K_1} \cup \bar{K}_n$ form a 3-parameter family of chromatically unique graphs; we have only to verify that the polynomial (2.7) is not divisible by $(\lambda - 1)^2$. This is easy for $k = 1$, $m \geq 4$, where (2.7) reduces to

$$P(\overline{K_2 \cup (m - 2)K_1}, \lambda) = (\lambda)_m + (\lambda)_{m-1} = (\lambda)_{m-2}(\lambda - m + 2)^2 \tag{4.18}$$

and for $k = 2$, $m \geq 4$, where (2.7) becomes

$$\begin{aligned} P(\overline{2K_2 \cup (m - 4)K_1}, \lambda) &= (\lambda)_m + 2(\lambda)_{m-1} + (\lambda)_{m-2} \\ &= (\lambda)_{m-2}\{\lambda^2 - (2m - 5)\lambda + m^2 - 5m + 7\} \end{aligned} \tag{4.19}$$

For greater values of k, it seems however better to use the formula (2.8) which shows (if we again replace $\underline{p}$ by $\underline{m}$) that we have only to prove that the polynomial $\Sigma_{i=0}^{k} \binom{k}{i} (\lambda - m + k)_{k-1}$ is not divisible by $(\lambda - 1)$, i.e., that this polynomial takes a value different from 0 for $\lambda = 1$:

$$\sum_{i=0}^{k} \binom{k}{i} (k + 1 - m)_{k-i} \neq 0 \tag{4.20}$$

But if we write out this sum (where it does not matter that its terms have alternating signs), we see that all its terms are divisible by $(m - k - 1)$, except the last one which becomes equal to 1. Hence the sum on the left hand side of (4.20) is a number of the form $1 + t(m - k - 1)$, and so indeed it cannot be zero since $m - k - 1 > 1$. (This follows from $m \geq 2k$ for $k > 2$.)

## 5. THE GRAPHS $K_n^+$, $C_n^+$ AND OTHER NEW FAMILIES OF CHROMATICALLY UNIQUE GRAPHS

As the special case $n = 1$ of the formula (4.15) we have the fact that the graph $K_m \cup K_1$, which is the complement of a star $K_{1,m}$, is chromatically unique (for $m = 1, 2, 3, \ldots$); its chromatic polynomial

$$P(\bar{K}_{1,m},\lambda) = \lambda^2(\lambda - 1)(\lambda - 2)\cdots(\lambda - m + 1) = (\lambda)_m \cdot \lambda \tag{5.1}$$

can be found in [6, Corollary 4.2.3].

One might ask what happens if we shift the exponent 2 to the right, i.e., if we consider the polynomial

$$\begin{aligned} P(G,\lambda) &= \lambda(\lambda - 1)\cdots(\lambda - \alpha + 1)(\lambda - \alpha)^2(\lambda - \alpha - 1)\cdots(\lambda - m + 1) \\ &= (\lambda)_m(\lambda - \alpha) \qquad \text{where } 1 \leq \alpha \leq m - 1 \end{aligned} \tag{5.2}$$

Of course, to justify the notation we must show that this is really the *chromatic* polynomial of a graph G (with $p = m + 1$ points). This is specially easy for $\alpha = 1$, where G turns out to be the graph which was called $K_m^+$ at the beginning of the foregoing section, and which might also be described as $\overline{K_{1,m-1} \cup K_1}$. In fact, it is readily checked (using Property IV) that

$$P(K_m^+,\lambda) = (\lambda - 1) \cdot P(K_m,\lambda) = (\lambda - 1) \cdot (\lambda)_m \tag{5.3}$$

In the general case ($2 \leq \alpha \leq m - 1$) it seems advisable to modify (5.2) as follows:

$$\begin{aligned} P(G,\lambda) &= \lambda(\lambda - 1)\cdots(\lambda - m + 1)\{(\lambda - m) + (m - \alpha)\} \\ &= \lambda(\lambda - 1)\cdots(\lambda - m) + (m - \alpha)\lambda(\lambda - 1)\cdots(\lambda - m + 1) \end{aligned}$$

i.e.,

$$P(G,\lambda) = (\lambda)_{m+1} + (m - \alpha)(\lambda)_m \tag{5.4}$$

It is now easy to find such a graph G whose chromatic polynomial is given by the right hand side of (5.4). In fact,

$$G = \overline{K_{1,m-\alpha} \cup \alpha K_1} \tag{5.5}$$

i.e., the complement of the union of a star $K_{1,m-\alpha}$ with $\alpha$ isolated points, is such a graph. This holds also in the case $\alpha = 1$, where we already mentioned above that in this case indeed

$$G = \overline{K_{1,m-1} \cup K_1} = K_m^+$$

To show that the graph (5.5) satisfies (5.4) we can use Property VII: The complement of $\overline{K_{1,m-\alpha} \cup \alpha K_1}$, i.e., the graph $K_{1,m-\alpha} \cup \alpha K_1$ has $m - \alpha$ lines, and because of (3.9) this counts for the coefficient of $(\lambda)_m$ in (5.4); and the coefficients of $(\lambda)_{m-1}$, $(\lambda)_{m-2}$, ..., $(\lambda)_1$ are zero because there are no subgraphs of the form (3.7) with $k < m$.

THEOREM 5.1. The graphs $\overline{K_{1,m-\alpha} \cup \alpha K_1}$ $(m \geq 2,\ 1 \leq \alpha \leq m - 1)$ whose chromatic polynomial is given by (5.2) or (5.4), are chromatically unique.

PROOF. We have to verify only the uniqueness. For this purpose let G be any graph satisfying (5.2) and (5.4); we will show that (5.5) holds. It follows from (5.4) and Property VII that the complement $\bar{G}$ has $p = m + 1$ points and $\bar{q} = m - \alpha$ lines, and since the coefficient of $(\lambda)_{m-1}$ in (5.4) is zero, $\bar{G}$ has neither a triangle $(K_3)$ nor two disjoint lines $(2K_2)$ as subgraphs.* Hence $\bar{G}$ cannot contain cycles, and so it must be a forest composed of $p - \bar{q} = \alpha + 1$ trees. But only one of these trees can contain lines (otherwise there would be again a subgraph isomorphic to $2K_2$); hence $\bar{G}$ must be the union of $\alpha$ isolated points and a tree with $\bar{q} = m - \alpha$ lines and $\bar{q} + 1 = m + 1 - \alpha$ points:

$$\bar{G} = T_{m+1-\alpha} \cup \alpha K_1 \tag{5.6}$$

But here the tree $T_{m+1-\alpha}$ must have diameter 2 (otherwise it would have at least one "forbidden" subgraph of type $2K_2$); in other words, $T_{m+1-\alpha}$ must be the star $K_{1,m-\alpha}$ (see [5, Exercise 4.15]). Hence we have proved that

$$\bar{G} = K_{1,m-\alpha} \cup \alpha K_1 \tag{5.7}$$

and this is equivalent to (5.5). ■

THEOREM 5.2. The graphs $\bar{K}_n \cup \overline{K_{1,m-\alpha} \cup \alpha K_1}$ $(m \geq 2,\ n \geq 1)$ are chromatically unique for $2 \leq \alpha \leq m - 1$.

---

*Otherwise $\bar{G}$ would have subgraphs of the form (3.7) with $k = p - 2$, leading to a *positive* coefficient of $(\lambda)_{m-1}$.

PROOF. Because of Theorem 5.1 and formula (5.2) the graphs $\overline{K_{1,m-\alpha} \cup \alpha K_1}$ satisfy the conditions for uniqueness stated in Theorem 4.1 whenever $2 \leq \alpha \leq m - 1$. ■

Returning to the case $\alpha = 1$ of Theorem 5.1, i.e., to the graph $K_m^+$ (see Figure 1 for $m = 3$), one might ask whether for other graphs H which are connected, point-symmetric (or vertex-transitive) and chromatically unique, is it true that the graph $H^+$ is chromatically unique too? We have not been able to decide this question in general; we could however decide it at least in the case that H is a cycle $C_n$:

THEOREM 5.3. The graphs $C_n^+$ $(n \geq 3)$ with $p = n + 1$ points are chromatically unique.

PROOF. Since the case $n = 3$ is covered by Theorem 5.1 for $m = 3$, $\alpha = 1$, we can suppose $n \geq 4$. (See Figure 3 for $n = 4$). From Property VI and formula (2.4), it follows that

$$P(C_n^+,\lambda) = (\lambda - 1)^2\left\{(\lambda - 1)^{n-1} + (-1)^n\right\} \tag{5.8}$$

Let G be any graph with the same chromatic polynomial, i.e.,

$$P(G,\lambda) = (\lambda - 1)^2\{(\lambda - 1)^{n-1} + (-1)^n\} \tag{5.9}$$

we have to show that G is isomorphic to $C_n^+$.

FIGURE 3. The graph $C_n^+$.

Now it follows from (5.8), (5.9), and Properties I and II that G has the same number of points, lines, and connected components as $C_n^+$; hence G also has $p = n + 1$ points, $q = n + 1$ lines, and is connected. But a connected graph in which the numbers of points and lines coincide is unicyclic (see [5, p. 41]), i.e., it has just one cycle $C_k$, where $3 \leq k \leq p$, to which points any number of trees might be attached. Hence, because of Property VI and formulas (1.1) and (2.4), the chromatic polynomial of G is

$$P(G,\lambda) = (\lambda - 1)^t\{(\lambda - 1)^{k-1} + (-1)^k\} \tag{5.10}$$

where

$$t = p - (k - 1) = n + 2 - k \tag{5.11}$$

since G has $p = n + 1$ points, and this number must be the degree of the polynomial (5.10) by Property I. On the other hand, by comparison of (5.10) with (5.9) it is seen that

$$t = 2, \qquad k = n \tag{5.12}$$

In other words, we have shown that the unique cycle $C_k$ in our graph G is a $C_n$, and since our graph G has $p = n + 1$ points, it is a $C_n$ at *one* of its points, a line (ending in a pendant point) has been attached; that is, G is isomorphic to $C_n^+$. Q.E.D.

## ACKNOWLEDGMENT

The author gratefully acknowledges enlightening discussions with Dr. Roberto Frucht, while the author was staying at Santa María University, Valparaíso, Chile, as visiting professor from December 1980 until May 1981.

## REFERENCES

1. C. Y. Chao and E. G. Whitehead, Jr., On chromatic equivalence of graphs, in *Theory and Applications of Graphs*, Y. Alavi and D. R. Lick, eds., Springer-Verlag, No. 642, May 1978.

2. C. Y. Chao and E. G. Whitehead, Jr., Chromatically unique graphs, Discrete Math. 27 (1979), 171-177.

3. R. W. Frucht, On a new method of computing chromatic polynomials of graphs, to appear in this volume.

4. R. E. Giudici and R. M. Vinke, A Table of Chromatic Polynomials, J. of Combinatorics, Information and System Sciences, Vol. 5, No. 4 (1980), 323-350.

5. F. Harary, *Graph Theory*, Addison-Wesley, 1969.

6. B. Loerinc, Computing chromatic polynomials for special families of graphs, Courant Computer Science Report No. 19 (February 1980), Courant Inst. of Math. Sciences, Computer Science Dept., New York Univ.

7. B. Loerinc, Chromatic uniqueness of the generalized θ-graph, Discrete Math. 23 (1978), 313-316.

8. R. C. Read, An introduction to chromatic polynomials, J. Combinatorial Theory 4 (1968), 52-71.

9. H. Whitney, A logical expansion in mathematics, Bull. Am. Math. Soc. 38 (1932), 572-579.

10. A. A. Zykov, On some properties of linear complexes, Am. Math. Soc. Transl. No. 79 (1952); translated from Math. Sb., 24, No. 66 (1949), 163-188.

# 10
# Nonparametric Solutions to the Variational Principle of Ideal Magnetohydrodynamics

OCTAVIO L. BETANCOURT* AND G. MCFADDEN† / Courant Institute of Mathematical Sciences, New York University, New York, New York

## 1. Introduction

In order to better understand magnetohydrodynamic equilibria in three dimensions, we study the lower dimensional cases. The solution of the three-dimensional problem is based upon the classical variational principle of ideal magnetohydrodynamics. The crucial assumption for the numerical method is the existence of a nested set of toroidal flux surfaces, which is then used as a coordinate. The question then arises of what happens when this assumption fails.

In this paper we study the nonparametric solutions to this variational problem in those cases where the direct solution is known to have islands.

In section 2 we describe a form of the variational principle for the slab geometry. In section 3 we analyze the one dimensional problem. In section 4 we discuss asymptotic expansions and numerical solutions to the two dimensional problem.

Current Affiliations:
*Computer Sciences Department, The City College of New York, New York, New York
†Mathematical Analysis Division, National Bureau of Standards, Gaithersburg, Maryland

## 2. The Variational Principle

The classical variational principle of magnetohydrodynamics [1] characterizes equilibria as stationary points of the potential energy

$$E = \iiint \left(\frac{1}{2} B^2 + \frac{p}{\gamma - 1}\right) dV \tag{1}$$

subject to appropriate constraints. Here B is the magnetic field, p is the fluid pressure related to the density $\rho$ by $p = \rho^{\gamma}$.

In [2], this variational principle has been reformulated in order to develop a numerical method to compute equilibria in three dimensions. The magnetic field B is represented by the product

$$B = \nabla s \times \nabla \phi \tag{2}$$

of two flux functions. In particular, s is assumed to be a single valued function whose level surfaces s = const. form a nested toroidal family of flux surfaces. It then follows that s can be used as a flux coordinate and that the periods of $\phi$ are the derivatives of given toroidal and poloidal fluxes. The mass within each flux surface is fixed, and the boundary is a given flux surface.

The Euler equations for this variational problem are

$$\nabla p = J \times B \, , \qquad J = \nabla \times B \tag{3}$$

where J is the current density. Together with the constraint $\nabla \cdot B = 0$ they correspond to the equations of magnetostatics.

In the simpler case of z independent equilibria, these reduce to the Grad-Shafranov equation [3],

$$- \Delta\psi = p'(\psi) + I I'(\psi) \tag{4}$$

where $\psi$ is the poloidal flux, and the pressure p and toroidal field I are functions of $\psi$ alone. The behavior of solutions to this equation is fairly well understood, and numerical methods to solve it are well developed.

Such is not the case for the full system of equations in three dimensions[4]. Solving the equation in toroidal geometry requires the existence of a single valued function solution of the equation

$$B \cdot \nabla s = 0 .$$

Moreover, the numerical method developed in [2] further assumes that the level surfaces of s are a family of nested toroidal flux surfaces. The KAM theory of ordinary differential equations shows that, in general, there will be regions in which the field lines of B are ergodic as well as others in which islands appear. Under certain conditions there also exist infinitely many nested toroidal surfaces.

We consider the simpler problem of using the variational principle to obtain nonparametric solutions in those cases in which islands appear in the solution of the Grad-Shafranov equation.

Let the magnetic field $B(x,y)$ be given by (3), periodic in x with period $2\pi$ and $p \equiv 0$. Also assume that there exists a family of flux surfaces with the nonparametric equation $y = y(x,s)$ for $-1 \leq s \leq 1$, and with given periodic boundary conditions $y(x,1) = Y_1(x)$ ; $y(x,-1) = Y_2(x)$. Using s as an independent variable, we can express the potential energy in the form

$$E = \frac{1}{2} \int_0^{2\pi} \int_{-1}^{1} \frac{f^2(1 + y_x^2) + \phi_x^2}{y_s} \, ds \, dx$$

where $\phi = g(s)x + \lambda(x,s) + f(s)z$, $\lambda$ is single valued and $f(s)$, $g(s)$ are the derivatives of given fluxes $\psi(s)$ and $G(s)$. The Euler equation for $\phi$ implies

$$I(s) = \frac{g(s)}{\int y_s \, dx}$$

which reduces the energy to

$$E = \frac{1}{2} \iint \frac{f^2(1 + y_x^2)}{y_s} \, ds \, dx + \frac{1}{2} \int \frac{g^2 \, ds}{\int y_s \, dx} \; . \tag{5}$$

The question becomes whether the problem of minimizing E among all functions y with prescribed boundary values has a nonparametric solution $y = y(x,s)$.

We can also consider the analog of the Grad-Shafranov equation. Given $I(\psi)$ we look for stationary points for the functional

$$\begin{aligned} H(\psi) &= \iint \left[ \frac{1}{2} (\nabla\psi)^2 - I^2(\psi) \right] dx \, dy \\ &= \iint \left[ \frac{f^2(1 + y_x^2)}{2y_s} - I^2(s) y_s \right] ds \, dx \end{aligned} \tag{6}$$

where $f(s) = \psi_s$ and $I(s)$ are prescribed, instead of f and g.

Note that when $I^2(\psi) = k_0^2 - \psi$, a solution to the Grad-Shafranov equation (4) is given by

$$\psi = \frac{1}{2} y^2 - \varepsilon \cosh y \cos x \tag{7}$$

where the boundary is defined by $\psi = \frac{1}{2}$. In this case, the island width is $2\sqrt{2\varepsilon}$ with the separatrix given by $\psi = \varepsilon$. Fig. 1 shows the flux surfaces for $\varepsilon = 0.01$.

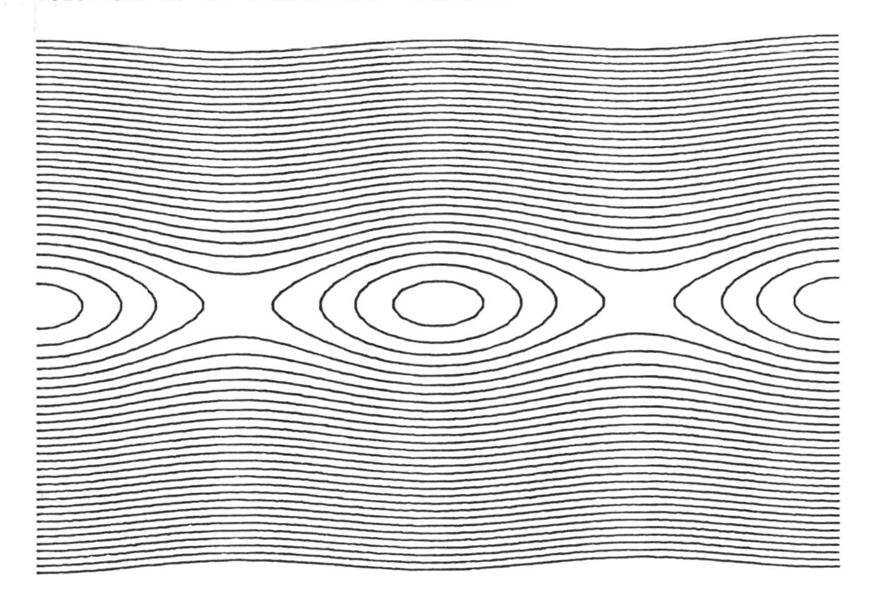

FIGURE 1. Flux surfaces psi (x,y) = constant for the Grad-Shafranov solution.

## 3. The one dimensional problem

Consider the one-dimensional nonparametric analog of the equation $\Delta\psi = 1$. Let $\psi = \frac{1}{2} s^2$ and look for a non-decreasing function y(s) which minimizes

$$\frac{1}{2} \int_{-1}^{1} \left(\frac{s^2}{y_s} + s^2 y_s\right) ds = \int \left[\frac{1}{2} (\nabla\psi)^2 + \psi\right] dy$$

with $y(1) = 1+\varepsilon$ ; $y(-1) = -1-\varepsilon$. The first variation results in

$$0 = \frac{1}{2} \int_{-1}^{1} \left(\frac{s^2}{y_s^2} - s^2\right)_s \delta y \, ds + \lim_{s\to 0^+} \frac{1}{2} \left(\frac{s^2}{y_s^2} - s^2\right) \delta y^+$$
$$- \lim_{s\to 0^-} \frac{1}{2} \left(\frac{s^2}{y_s^2} - s^2\right) \delta y^- ,$$

where a possible discontinuity of y is allowed at the resonance value s = 0. In the region s > 0, the Euler equation has the general solution

$$y(s) = \sqrt{s^2 + a^2} + b \, .$$

The natural boundary condition at $s = 0$, $\lim_{s \to 0^+} \frac{s^2}{2y_s^2} = 0$ implies $a = 0$; and the remaining condition at $s = 1$ gives $y = s+\varepsilon$. For $\varepsilon > 0$, the solution with $y(-s) = -y(s)$ is monotone with a jump at $s = 0$ equal to $2\varepsilon$.

For $\varepsilon < 0$, however, this solution is not monotonic at $s = 0$. The boundary condition at $s = 1$ requires $b = 1+\varepsilon-\sqrt{1+a^2}$, and minimizing the functional subject to the monotonicity condition $y(0) \geq 0$ implies $y(0) = 0$. These results are shown in Figure 2.

Consider the solution $\psi(y)$ for the case $\varepsilon > 0$. We have $\Delta\psi = 1$ for $|y| > \varepsilon$ and $\psi_y = 0$ at $|y| = \varepsilon$. As a result, we cannot continue the solution to the region $|y| < \varepsilon$ such that both $\psi_y$ and $\psi_{yy}$ be continuous at $|y| = \varepsilon$ and $\Delta\psi = 1$ for $|y| < \varepsilon$.

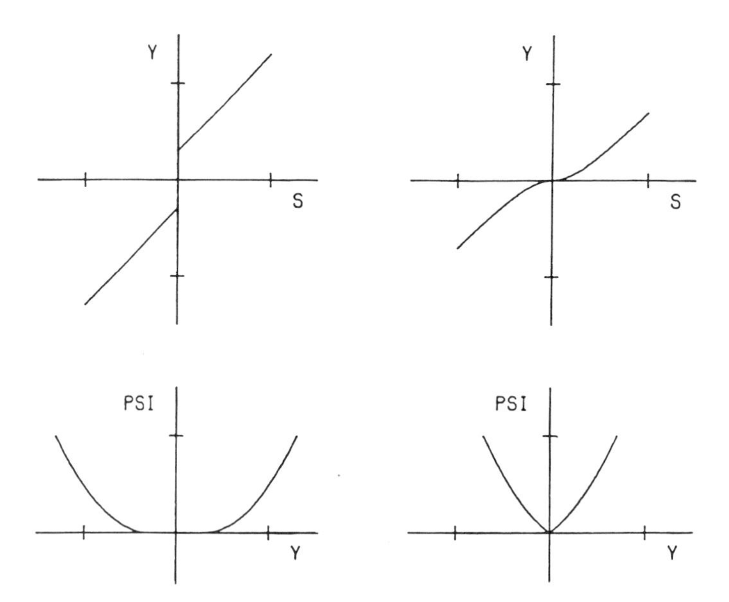


FIGURE 2. The one-dimensional variational principle. The left side is $\varepsilon > 0$. The right side is $\varepsilon < 0$.

If we want to obtain a smooth solution, we must determine the value of the jump of y at s = 0 so that the outer solution for $s \geq 0$ matches a solution of $\Delta\psi = 1$ for $|y| < |y(0)|$. Therefore, we would require that y(0) be such that

$$y = \psi_y = \lim_{s \to 0^+} \frac{\psi_s}{y_s} \quad ;$$

which implies b = 0. The solution is then given by

$$\psi = \frac{1}{2} s^2 = \frac{1}{2} (y^2 - 2\varepsilon - \varepsilon^2) = \frac{1}{2} (y^2 - y^2(0)) \ .$$

Note that this gives a jump of order $\sqrt{\varepsilon}$.

In the case of given fluxes $f(s) = \frac{1}{2} s^2$ and $g(s) = \sqrt{k^2 - s^2}$ , the functional is given by

$$E(y) = \frac{1}{2} \int_{-1}^{1} \left( \frac{s^2}{y_s} + \frac{k^2 - s^2}{y_s} \right) ds = \frac{1}{2} \int_{-1}^{1} \frac{k^2}{y_s} ds \ ,$$

and the minimizing nondecreasing solution is $y = (1+\varepsilon)s$. In this case no discontinuous solutions occur.

## 4. The two dimensional problem

The numerical method is based upon minimizing the functionals (5) and (6) by introducing the paths of steepest descent

$$a\, y_{tt} + e\, y_t = - L(y) \tag{8}$$

where the first variation of the functional is $\int L(y)\, \delta y$ . A discrete approximation to L(y) is obtained by writing a second order accurate approximation to the functional, and differentiating explicitly with respect to the values of the solution $y_{ij}$ at the nodes. This results in an approximation in conservation form, which allows for an adequate numerical treatment of weak solutions.

The resulting equations are solved by an iteration procedure suggested by a discrete approximation to the time dependent equation (8). The coefficient a is chosen to satisfy the Courant-Friedrichs-Lewy condition and the descent coefficient e is chosen to maximize the convergence rate [2].

Figure 3 shows the flux surfaces $y = y(s_0, x)$ resulting from the numerical solution of the nonparametric variational problem (6) for the equation $\Delta\psi = 1$, $\psi(s) = s^2/2$. The island is clearly visible as a discontinuity at the resonance surface $s = 0$. The island width, defined to be the maximum jump of y at $s = 0$ is of order $\varepsilon$, the amplitude of the perturbation of the outer surface. Moreover, the islands are connected by straight "slits" rather

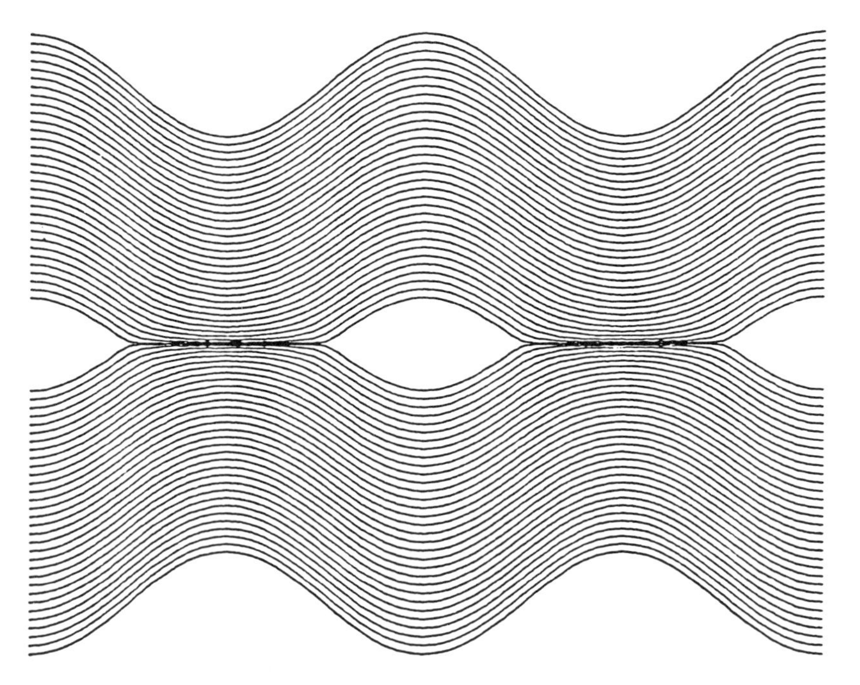

FIGURE 3. Flux surfaces for the case of fixed current.

than through x points. In addition, the computed current density $\Delta\psi(s)$ becomes singular at $s = 0$, while $\Delta\psi(s) = 1$ for $s \neq 0$. A similar type of island has been described in [5].

The form of the singularity and a formula for the island width can be obtained through an asymptotic expansion for the solution in terms of the perturbation amplitude $\varepsilon$. The Euler equation for the functional (6) is

$$\left(\frac{s^2(1+y_x^2)}{2y_s^2}\right)_s - \left(\frac{s^2 y_x}{y_s}\right)_x - s = 0$$

with boundary conditions $y(x,1) = 1 + \varepsilon \cos x$ ; and $y(x,-s) = -y(x,s)$. The solution for $\varepsilon = 0$ is $y = s$, and a formal linearization $y = s + \varepsilon Y(s,u)$ results in

$$(s^2 Y_s)_s + s^2 Y_{xx} = 0$$

with $Y(x,1) = \cos x$. A general solution is

$$Y(s,x) = \frac{(A \sinh s + B \cosh s) \cos x}{s} .$$

Consider the case $B = 0$, so that

$$y = s + \bar{\varepsilon} \cos x \frac{\sinh s}{s} \quad \text{with} \quad \bar{\varepsilon} = \varepsilon / \sinh 1 .$$

This solution is adequate for $s \sim 1$, but it breaks down when $s \sim \varepsilon$, since then both terms are of the same order of magnitude. This suggests the rescaling $s = \varepsilon\tau$, $y = \varepsilon v$ for an inner expansion. The resulting equation for $v(\tau,x)$ to lower order in $\varepsilon$ is

$$\left(\frac{\tau^2}{2v_\tau^2}\right)_\tau - \tau = 0$$

with general solution given by

$$v(x,\tau) = \sqrt{a^2(x) + \tau^2} + b(x) .$$

The functions a(x) and b(x) are determined by requiring that as $\tau \to \infty$ the inner solution should match the outer solution as $s \to 0$, and that they satisfy the proper conditions at $\tau = 0$.

Following the one dimensional case, for cos x > 0 we require that $\lim_{s\to 0} s^2(1+ y_x^2) / 2y_s^2 = 0$. This implies $a^2(x) = 0$, and the boundary condition for $\tau \to \infty$ gives $b(x) = \cos x$. When cos x < 0 this solution does not satisfy the monotonicity requirements at s = 0. Therefore, we require $v(x,0) = 0$ which results in $v(x,\tau) = \sqrt{\tau^2 + \cos^2 x} + \cos x$.

Finally, the outer and inner solutions can be combined into a composite expansion for $s \geq 0$,

$$y = \begin{cases} \dfrac{\bar{\varepsilon} \cos x \sinh s}{s} + s & \text{if} \quad \cos x > 0 \\ \dfrac{\bar{\varepsilon} \cos x \sinh s}{s} + \sqrt{s^2+\bar{\varepsilon}^2\cos^2 x} & \text{if} \quad \cos x < 0 \end{cases}$$

and $y(x,s) = -y(x,-s)$ for s < 0. Results from this expansion agree extremely well with the numerical results. Furthermore it shows that the island width is $2\bar{\varepsilon}$.

For cos x < 0, the expression can be inverted to give s(x,y) and

$$\psi = s^2 / 2 = y^2/2 - \bar{\varepsilon} \cos x \sinh|y|$$

so that $\Delta\psi = 1 - 2\bar{\varepsilon} \cos x \, \delta(y)$, where $\delta(y)$ is the delta function.

This example shows that the assumption of nested flux surfaces need not rule out the occurrence of islands. It should be noted that the island width is of the order $\varepsilon$

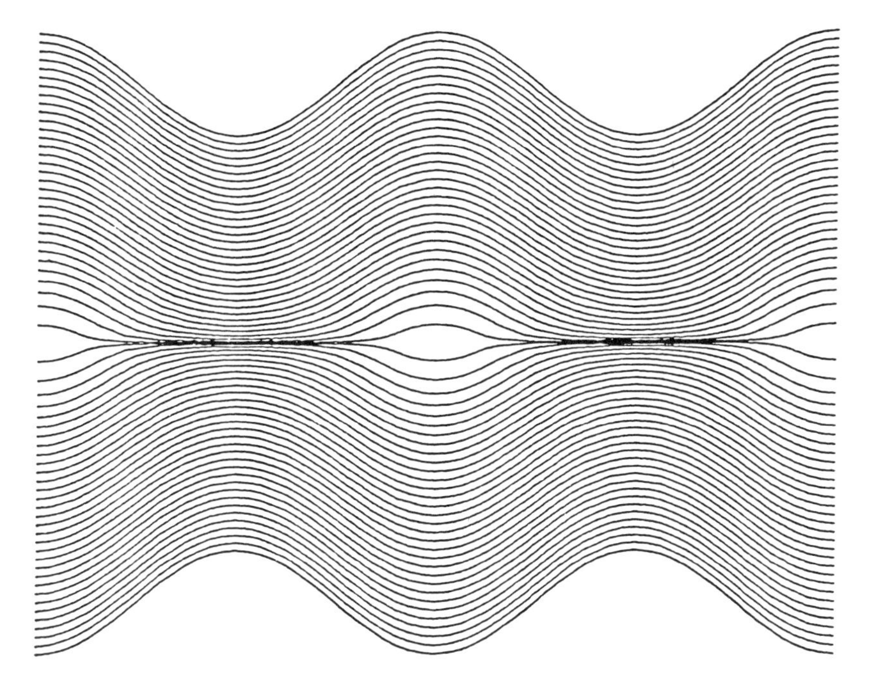

FIGURE 4. Flux surfaces for the case of fixed fluxes.

and is accompanied by a singular current sheet, rather than the more familiar $\sqrt{\varepsilon}$ dependence in (7).

Figure 4 shows the flux surfaces $y = y(s_0,x)$ for the numerical solution that minimizes the functional given by (5). The fluxes f and g are chosen such that $y = s$ and $\Delta\psi = 1$ for $\varepsilon = 0$. The amplitude of the perturbation $\varepsilon$ of the outer wall is the same as in Figure 3. Although the solutions are quite similar away from the resonant surface $s = 0$, the island width is much smaller. In fact, it is found that as the mesh size h is refined, the island width converges to zero as $\sqrt{h}$.

The computed values of the toroidal current density $\Delta\psi$ also show a singular current sheet at $s = 0$. In addition, there is a small transition layer $|s| < \eta(\varepsilon)$ in which $\Delta\psi(s)$ decreases from one to zero. The behavior of the current density near the resonant surface may be used to estimate the island width that would be obtained under a different set of constraints [4].

As we have seen the assumption of nested surfaces results in singular currents at the resonant surface. We would like to compute a solution which instead models the island by a region of constant current density and constant pressure. To achieve this we must find the location of the separatrix between the island region and the outer region such that the field B and the current density $\Delta\psi$ are continuous across it.

This results in a free boundary problem for the matching of the solution $\Delta\psi = k$, $p = \text{const.}$ with the outer solution at $s = s_0$. The precise algorithm is currently under investigation.

## Acknowledgments

This work was supported by NASA Grant NS6-1579 and by the U. S. Department of Energy under contract DE-AC02-76ER0-3077.

## References

[1] Kruskal, M. D. and Kulsrud, R. M., "Equilibrium of a magnetically confined plasma in a torus," Phys. Fluids 1, 265-274 (1958).

[2] Bauer, F., Betancourt, O. and Garabedian, P., "A Computational Method in Plasma Physics," Springer Series in Computational Physics, Springer Verlag, 1978.

[3] Bateman, G., MHD Instabilities (MIT Press, Cambridge, 1978).

[4] Betancourt, O. and Garabedian, P., "Numerical Analysis of Equilibria with Islands in Magnetohydrodynamics," to appear.

[5] Hu, P. N., "Reconnection of Magnetic Lines," Proceedings of the 23rd Annual Meeting of the Division of Plasma Physics in New York, October 1981.

# 11
# A Metric for a Flag Space

WILFRED REYES* / Instituto de Matemática, Universidade Estadual de Campinas, Campinas, São Paulo, Brazil

## 1. The real flag manifold as a homogeneous space

Let $\mathbb{R}^n$ be provided with its canonical inner product $\langle\ ,\rangle$, and let $n = n_1 + \ldots + n_k$, $n_i > 0$. Then to each sequence of subspaces $(V^1, \ldots, V^k)$, $\dim V^i = n_i$, such that $V^1 \oplus \ldots \oplus V^k = \mathbb{R}^n$ (orthogonal sum with respect to $\langle\ ,\rangle$), we associate a flag

$$0 \subset V^1 \subset V^1 \oplus V^2 \subset \ldots \subset V^1 \oplus \ldots \oplus V^k = \mathbb{R}^n .$$

Conversely, to any flag

$$0 \subset W^1 \subset W^2 \subset \ldots \subset W^k = R^n , \quad \dim W^i = n_i ,$$

there corresponds an orthogonal sequence

$$(W^1, W^{1,2}, \ldots, W^{k-1,k}), \quad W^{i-1} \oplus W^{i-1,i} = W^i .$$

We note by $F(n_1, \ldots, n_k)$ the set of orthogonal sequences as defined above and we state their properties in the following

*Current Affiliation: Departamento de Matemáticas, Facultad de Ciencias, Básicas y Farmacéuticas, Universidad de Chile, Santiago, Chile

Proposition 1. The orthogonal group $O(n)$ acts transitively on $F(n_1, \ldots, n_k)$. The isotropy group $H = H(n_1, \ldots, n_k)$ at a point is diffeomorphic to the direct product $O(n_1) x \ldots x O(n_k)$ ; thus by means of the identificacion

$$F(n_1, \ldots, n_k) = O(n)/H(n_1, \ldots, n_k)$$

$F(n_1, \ldots, n_k)$ becomes a homogeneous space of dimension $(n^2 - n_1^2 - \ldots - n_k^2)/2$ .

The former identification induces that of the tangent space of $O(n)/H$ at $H$ with $m = o(n)\setminus g$, where $o(n)$ is the Lie-algebra of $O(n)$ and $g = g(n_1, \ldots, n_k) \approx o(n_1) + \ldots + o(n_k)$ , that of $H$ .

If $m$ is provided with the norm

$$\|X\|^2 = \mathrm{tr}\, XX'/2 ,$$

$X'$ denoting the transpose of $X$ , we have an invariant riemannian metric on $O(n)/H$ at $H$ , and consequently, the space being homogeneous, the metric is everywhere defined. // (See [1] or [2]).

To be more explicit, take a differentiable curve on $O(n)$:

$f : (-\varepsilon,\varepsilon) \to O(n)$ , $t \to f(t) = (f_1(t), \ldots, f_n(t))$ , construct the following matrix in $o(n)$ :

$$\langle f,f^{\cdot} \rangle = f'(0)f^{\cdot}(0) = (f_i'f_j^{\cdot})_{t=0} = (\langle f_i,f_j^{\cdot} \rangle)$$

and consider next the projection of the matrix on $m$ :

$$\langle f,f^{\cdot} \rangle_m = (\langle f_{i_r},f^{\cdot}_{j_s} \rangle)_{r\neq s} , \quad n_1 + \ldots + n_{r-1} \leq i_r \leq n_1 + \ldots + n_r .$$

Then the expression for the metric becomes

$$\|\langle f, f^{\cdot} \rangle_m \|^2 = \sum_{r<s} \langle \dot{f}_{i_r}, f_{j_s} \rangle^2$$

where the summation extends over all indices $i_r$, $j_s$ and where the point means differentiation. One verifies easily that the metric is well defined. We use it as a metric for the respective flag space.

## 2. A system of local coordinates for double flags

For simplicity we state the following ranges of indices:

$\alpha, \beta, \rho, \sigma = 1, \ldots, p$ ; $a, b, r, s = p + 1, \ldots, q$ ;

$A, B, R, S = q + 1, \ldots, n$ and we adhere to the summation convention on repeated indices.

Given the double flag $0 \subset V_o^p \subset V_o^q \subset \mathbb{R}^n$, we note the decomposition it induces on $\mathbb{R}^n$ by $V_o^p \oplus V_o^{q-p} \oplus V_o^{n-q}$. Now, by elementary algebra, $R^n$ possesses an orthonormal basis $(e_i)$ such that $[e_\alpha] = V_o^p$, $[e_a] = V_o^{q-p}$, $[e_A] = V_o^{n-q}$ where the brackets denote the span of the vectors within.

By a neighbourhood of the double flag above we mean the set of all double flags $0 \subset V^p \subset V^q \subset \mathbb{R}^n$ such that the following two conditions are satisfied:

(1) the projection of $V^p$ onto $V_o^p$ parallel to the orthogonal complement $V_o^{n-p}$ is non degenerate;

(2) the projection of $V^q$ onto $V_o^q$ parallel to $V_o^{n-q}$ is also non degenerate.

In other words we have the following

Proposition 2. There exists a basis $(\varepsilon_i)$ of $\mathbb{R}^n$ such that $V^p = [\varepsilon_\alpha]$, $V^q = [\varepsilon_\alpha, \varepsilon_a]$.

In terms of the orthonormal basis $(e_i)$ above, the $\varepsilon_i$ are of the form

$$(3) \quad \varepsilon_\alpha = e_\alpha + x^a_\alpha e_a + z^A_\alpha e_A \,, \quad \varepsilon_a = e_a + y^A_a e_A \,.$$

Moreover, we have also a basis for $V^{n-q}$, namely:

$$(4) \quad \varepsilon_A = z^\alpha_A e_\alpha + y^a_A e_a + e_A \,, \quad y^a_A = -y^A_a \,, \quad z^\alpha_A = x^a_\alpha y^A_a - z^A_\alpha \,.$$

Thus to the flag $0 \subset V^p \subset V^q \subset R^n$ we associate a point $(x^a_\alpha, y^A_a, z^A_\alpha)$ in $\mathbb{R}^{p(n-q)+(q-p)(n-q)}$ ; conversely, in view of (3) to any such a point there corresponds a flag near $0 \subset V^p_o \subset V^q_o \subset \mathbb{R}^n$//.

Perfom a change of coordinates in $\mathbb{R}^n$ :

$$(5) \quad m : e_i \to \bar{e}_i = m^j_i e_j \,, \qquad i,j = 1, \ldots, n \,.$$

So that relations (3) become:

$$(6) \quad \bar{\varepsilon}_\alpha = \bar{e}_\alpha + \bar{x}^a_\alpha \bar{e}_a + \bar{z}^A_\alpha \bar{e}_A \,, \quad \bar{\varepsilon}_a = \bar{e}_a + \bar{y}^A_a \bar{e}_A \,.$$

If $m$ leaves $0 \subset V^p \subset V^q \subset R^n$ invariant there is a set of constants $c^\beta_\alpha$, $c^\beta_a$, $c^b_a$ , such that:

$$(7) \quad \bar{\varepsilon}_\alpha = c^\beta_\alpha \varepsilon_\beta \,, \quad \bar{\varepsilon}_a = c^\beta_a \varepsilon_\beta + c^b_a \varepsilon_b \,.$$

Substituting in (7) relations (3), (5) and (6) we eliminate those constants and state the resulting coordinate transformation in $F(p, q - p, n - q)$ as:

$$(8) \quad (\bar{x}^a_\alpha, \bar{y}^A_a, \bar{z}^A_\alpha) \to (x^a_\alpha, y^A_a, z^A_\alpha) \,,$$

or, in terms of the transformation $m$ in (5):

$$(9)\quad m^b_\alpha + m^{b-a}_a x^a_\alpha + m^{b-A}_A z^A_\alpha = (m^\beta_\alpha + m^{\beta-a}_a x^a_\alpha + m^{\beta-A}_A z^A_\alpha)\, x^b_\beta ,$$

$$m^B_\alpha + m^{B-a}_a x^a_\alpha + m^{B-A}_A z^A_\alpha = (m^\beta_\alpha + m^{\beta-a}_a x^a_\alpha + m^{\beta-A}_A z^A_\alpha)\, z^B_\beta ,$$

$$m^B_a m^\beta_\alpha - m^B_\alpha m^\beta_a + (m^B_A m^\beta_\alpha - m^B_\alpha m^\beta_A)\, \bar{y}^A_a + (m^B_A m^\beta_a - m^B_a m^\beta_A)\, \bar{z}^\alpha_A =$$

$$((m^b_a m^\beta_\alpha - m^b_\alpha m^\beta_a + (m^b_A m^\beta_\alpha - m^b_\alpha m^\beta_A)\, \bar{y}^A_a + (m^b_A m^\beta_a - m^b_a m^\beta_A)\, \bar{z}^\alpha_A\,)\, j^\beta_b$$

So we have a further property in

<u>Proposition 3</u>. $F(p,\, q-p,\, n-q)$ is a rational manifold.//

## 3. The metric in coordinates

Relations 2 - (3), -(4) imply

$$(1)\quad g_{\alpha\beta} = \langle \varepsilon_\alpha, \varepsilon_\beta \rangle = \delta_{\alpha\beta} + x^a_\alpha x^a_\beta + z^A_\alpha z^A_\beta ,$$

$$g_{ab} = \langle \varepsilon_a, \varepsilon_b \rangle = \delta_{ab} + y^A_a y^A_b ,$$

$$g_{AB} = \langle \varepsilon_A, \varepsilon_B \rangle = \delta_{AB} + y^a_A y^a_B + z^\alpha_A z^\alpha_B ,$$

$$g_{\alpha A} = \langle \varepsilon_\alpha, \varepsilon_A \rangle = g_{aA} = \langle \varepsilon_a, \varepsilon_A \rangle = 0$$

We may also suppose

$$g_{\alpha a} = \langle \varepsilon_\alpha, \varepsilon_a \rangle = x^a_\alpha + y^A_a z^A_\alpha = 0 .$$

Otherwise we restrict ourselves if necessary to a neighbourhood of $0 \subset V^p_o \subset V^q_o \subset \mathbb{R}^n$ such that for any flag within the neighbourhood the corresponding matrix $(g_{\alpha\beta})$ possesses an inverse $(g^{\alpha\beta})$, and we proceed to perform in $V^q = [\varepsilon_\alpha, \varepsilon_a]$ a change of basis:

$$(2)\quad (\varepsilon_\alpha, \varepsilon_a) \to (\varepsilon_\alpha, \varepsilon_a - d^\alpha_a \varepsilon_\alpha) ,\quad d^\alpha_a = g_{a\beta} g^{\beta\alpha} .$$

Moreover, by the Gramm-Schmidt algorithm there exists an ortho-normal basis of $\mathbb{R}^n$, say $(f_\alpha, f_a, f_A)$, such that:

$$(3) \quad [f_\alpha] = [\varepsilon_\alpha] = V^p , \quad [f_a] = [\varepsilon_a] = V^{q-p} , \quad [f_A] = [\varepsilon_A] = V^{n-q}$$

The former relations imply the following:

$$(4) \quad f_\alpha = L^\rho_\alpha \varepsilon_\rho , \quad f_a = M^r_a \varepsilon_r , \quad f_A = N^R_A \varepsilon_R$$

for a certain set of coefficients $\{L^\rho_\alpha, M^r_a, N^R_A\}$.

Further:

$$(5) \quad L^\rho_\alpha L^\sigma_\beta g_{\rho\sigma} = \delta_{\alpha\beta} , \quad M^r_a M^s_b g_{rs} = \delta_{ab} , \quad N^R_A N^S_B g_{RS} = \delta_{RS} .$$

These relations, as it is well known, imply:

$$(6) \quad g^{\rho\sigma} = L^\rho_\alpha L^\sigma_\alpha , \quad g^{rs} = M^r_a M^s_a , \quad g^{RS} = N^R_A N^S_A ,$$

whithin a neighbourhood of $0 \subset V^p_o \subset V^q_o \subset \mathbb{R}^n$ where the matrices $(g_{\rho\sigma})$, $(g_{rs})$ and $(g_{RS})$ are invertible.

Formulae 2 - (3), -(4) yield also:

$$(7) \quad \langle d\varepsilon_\rho, \varepsilon_r \rangle = \langle dx^a_\rho e_a + dz^A_\rho e_A , \quad e_r + y^B_r e_B \rangle =$$

$$dx^r_\rho + y^A_r dz^A_\rho := \omega_{\rho r} ,$$

$$\langle d\varepsilon_\rho, \varepsilon_R \rangle = dz^R_\rho + y^a_R dx^a_\rho := \omega_{\rho R} ,$$

$$\langle d\varepsilon_r, \varepsilon_R \rangle = dy^R_r := \omega_{rR} .$$

Now, the metric given in Section 1 becomes, in terms of the ortho-normal basis $(f_\alpha, f_a, f_A)$ above:

$$(8) \quad ds^2(f) = \sum_{\alpha,a} \langle df_\alpha, f_a \rangle^2 + \sum_{\alpha,A} \langle df_\alpha, f_A \rangle^2 + \sum_{a,A} \langle df_a, f_A \rangle^2$$

But this metric can be formulated also in terms of the basis $(\varepsilon_\alpha, \varepsilon_a, \varepsilon_A)$ .

In fact, proceed to compute the various summations: from (4) we have:

$$(9) \qquad \langle df_\alpha, f_a \rangle = L_\alpha^\rho \langle d\varepsilon_\rho, \varepsilon_r \rangle M_a^r = L_\alpha^\rho \omega_{\rho r} M_a^r ,$$

because of the supposed relation $\langle \varepsilon_\rho, \varepsilon_r \rangle = g_{\rho r} = 0$

So we begin to work in formula (8),

$$(10) \quad \langle df_\alpha, f_a \rangle^2 = L_\alpha^\rho \omega_{r\rho} M_a^r M_a^s \omega_{\sigma s} L_\alpha^\sigma = L_\alpha^\rho \omega_{r\rho} g^{rs} \omega_{\sigma s} L_\alpha^\sigma = g^{\sigma\rho} \omega_{r\rho} g^{rs} \omega_{\sigma s} ,$$

because of relations (6). A similar calculation gives

$$(11) \quad \langle df_\alpha, f_A \rangle^2 = g^{\sigma\rho} \omega_{R\rho} g^{RS} \omega_{\sigma S} ,$$

$$\langle df_a, f_A \rangle^2 = g^{sr} \omega_{Rr} g^{RS} \omega_{sS} .$$

Substituting relations (10) and (11) in formula (8) we obtain the metric for the double flag $0 \subset V^p \subset \dot{V}^q \subset \mathbb{R}^n$ in classical deguise

$$(12) \quad ds^2 = g^{\sigma\rho} \omega_{r\rho} g^{rs} \omega_{\sigma s} + g^{\sigma\rho} \omega_{R\rho} g^{RS} \omega_{\sigma S} + g^{sr} \omega_{Rr} g^{RS} \omega_{sS}$$

We state the former conclusion as a

Theorem. The riemannian metric for a double flag $0 \subset V^p \subset V^q \subset \mathbb{R}^n$ can be formulated in terms of the local coordinates stated in Section 2.

Comments.

If $\alpha = 1, \ldots, q$ , $A = q + 1, \ldots, n$ in 2 - (3), -(4) these equations become:

$$\varepsilon_\alpha = e_\alpha + z_\alpha^A e_A \ , \quad \varepsilon_A = z_A^\alpha e_\alpha + e_A$$

and the flag manifold $[e_\alpha]$ is the Grassmann manifold of q-planes in $\mathbb{R}^n$. In this case the metric 3 - (12) simplifies to:

$$ds^2 = g^{\sigma\rho}\omega_{R\rho} g^{RS}\omega_{\sigma S} \ .$$

Set $z = (z_\alpha^A)$, then $(g^{\sigma\rho}) = (1 + z'z)^{-1}$, $(g^{RS}) = (1 + zz')^{-1}$,

$$(\omega_{R\rho}) = (dz_R^\rho) = dz' \ , \quad (\omega_{\rho R}) = (dz_\rho^R) = dz$$

So the metric may be writen in an intrinsec fashion as:

$$ds^2 = \mathrm{Trace}(I + z'z)^{-1} dz'(1 + zz')^{-1} dz \ .$$

See for details on the later [3] and [4] .

In reference [5] we include some computations of the Christoffel symbols for the flag manifold as well as some geometric interpretation that we hope to pursue further.

## References

[1] Kobayashi-Nomizu: Foundations of Differential Geometry, Volume 2. 1969.

[2] Hangan, Th.: Sur la géométrie différentielle des variétés de drapeaux. Symposia Mathematica, Vol X, 1971.

[3] Hangan, Th.: Thèse d'Etat, Mulhouse, 1979.

[4] Wong, Y.C.: Differential Geometry of Grassmann manifolds. Proceedings of the N.Ac. of Sc., U.S.A. 1967.

[5] Reyes, W.: Thèse 3e. c. Strasbourg, 1981.

# 12
# Domains of Holomorphy in Banach Spaces

JORGE MUJICA / Instituto de Matemática, Estatística e Ciência da Computação, Universidade Estadual de Campinas, Campinas, São Paulo, Brazil

## INTRODUCTION

In 1911 Levi [7] posed the problem of whether every pseudoconvex domain in $\mathbb{C}^n$ is a domain of holomorphy. In 1942 Oka [14] solved the problem in the affirmative in the case $n = 2$, and it was only in 1953-1954 that Oka [15], Bremermann [1] and Norguet [10] independently solved the problem for every $n \geq 2$.

In 1937 Oka [13] obtained an approximation theorem for functions holomorphic on a neighborhood of a holomorphically convex compact subset of a domain of holomorphy. Weil [16] had previously obtained a similar result for a smaller class of domains.

During the last fifteen years this kind of results has been studied within the framework of Banach spaces. In 1972 Gruman and Kiselman [3] solved the Levi problem in Banach spaces with a Schauder basis, proving that all pseudoconvex domains in such spaces are domains of existence. A similar result had been previously obtained by Dineen and Hirschowitz [2] under an additional hypothesis on the domain. Then Noverraz [11] extended the theorem of Gruman and Kiselman to the case of Banach spaces with the bounded approximation property, and next he obtained an approximation theorem of the Oka-Weil type in such spaces [12].

This paper is mostly a survey of the main results on domains of holomorphy in Banach spaces, with special emphasis on the case of Banach space with a Schauder basis. The core of this paper are Sections 3 and 4. In

Section 3 we give a new proof of the theorem of Gruman and Kiselman [3], and in Section 4 we prove an approximation theorem of the Oka-Weil type, which sharpens the previous result of Noverraz [12].

## 1. NOTATION AND TERMINOLOGY

The letter E will always represent a complex Banach space, and the letters U, V, and W will represent nonvoid open subsets of E.

We will denote by H(U) the collection of all complex-valued holomorphic functions on U. We recall that a function $f: U \to \mathbb{C}$ is *holomorphic* if it is continuous and its restriction to each complex line is holomorphic in the classical sense. If $f \in H(U)$ and $x \in U$ then f admits a series expansion

$$f(x + s) = \sum_{n=0}^{\infty} f^{(n)}(x)(s)$$

where $f^{(n)}(x)$ is a continuous n-homogeneous polynomial on E. If A is an arbitrary subset of E (not necessarily open) then H(A) will denote the collection of all complex-valued functions which are holomorphic on some open neighborhood of A.

We will represent by Ps(U) the collection of all *plurisubharmonic* functions on U. We recall that a function $f: U \to [-\infty,+\infty)$ is *plurisubharmonic* if it is upper semicontinuous and its restriction to each complex line is subharmonic in the classical sense. Finally we will represent by Psc(U) the collection of all plurisubharmonic functions on U which are continuous.

We refer to the books of Hörmander [5] and Noverraz [11] for the theory of holomorphic and plurisubharmonic functions on finite dimensions and infinite dimensions, respectively.

We shall be dealing with Banach spaces with a Schauder basis and we refer to the book of Lindenstrauss and Tzafriri [8] for the properties of these spaces.

## 2. DOMAINS OF HOLOMORPHY IN BANACH SPACES

2.1 DEFINITION. An open set $U \subset E$ is said to be a *domain of holomorphy* if there are no open sets $U_1$ and $U_2$ with the following properties:

(a) $U_1 \not\subset U$ and $U_1$ is connected.

(b) $\emptyset \neq U_2 \subset U \cap U_1$.

(c) For each $f \in H(U)$ there is $f_1 \in H(U_1)$ such that $f_1 = f$ on $U_2$.

2.2 DEFINITION. An open set $U \subset E$ is said to be a *domain of existence* for a function $f \in H(U)$ if there are no open sets $U_1$ and $U_2$ and a function $f_1 \in H(U_1)$ with the following properties:

(a) $U_1 \not\subset U$ and $U_1$ is connected.
(b) $\emptyset \neq U_2 \subset U \cap U_1$.
(c) $f_1 = f$ on $U_2$.

2.3 DEFINITION. Let U be an open set in E. For each set $A \subset U$ we define its $H(U)$-*hull* by

$$\hat{A}_{H(U)} = \{x \in U : |f(x)| \leq \sup_A |f| \text{ for all } f \in H(U)\}$$

The open set U is said to be $H(U)$-*convex* or *holomorphically convex* if $d(\hat{K}_{H(U)};E\backslash U) > 0$ for each compact set $K \subset U$.

2.4 DEFINITION. Let U be an open set in E. For each set $A \subset U$ we define its $Psc(U)$-*hull* by

$$\hat{A}_{Psc(U)} = \{x \in U : f(x) \leq \sup_A f \quad \text{for all } f \in Psc(U)\}$$

The open set U is said to be $Psc(U)$-*convex* if $d(K_{Psc(U)};E\backslash U) > 0$ for each compact set $K \subset U$.

2.5 DEFINITION. An open set $U \subset E$ is said to be *pseudoconvex* if the function $-\log d(x;E\backslash U)$ is plurisubharmonic on U.

2.6 THEOREM. [4] For an open set $U \subset E$ consider the following conditions:

(a) U is a domain of existence.
(b) U is a domain of holomorphy.
(c) U is $H(U)$-convex.
(d) U is $Psc(U)$-convex.
(e) U is pseudoconvex.

Then (a) $\Rightarrow$ (b) $\Rightarrow$ (c) $\Rightarrow$ (d) $\Leftrightarrow$ (e).

PROOF. The implications (a) $\Rightarrow$ (b), (c) $\Rightarrow$ (d), and (e) $\Rightarrow$ (d) are clear. The implication (d) $\Rightarrow$ (e) follows at once from the corresponding implication in finite dimensions; see, e.g., Hörmander [5, Theorem 2.6.7]. Thus we only have to prove the implication (b) $\Rightarrow$ (c).

Consider a compact set $K \subset U$. Let $y \in \hat{K}_{H(U)}$ and let $0 < \delta < d(K;E\setminus U)$. We will show that for each $f \in H(U)$ there exists $f_1 \in H(B(y;\delta))$ such that $f_1 = f$ on a neighborhood of $y$. Since $U$ is a domain of holomorphy, this will imply that $B(Y;\delta) \subset U$ and hence (c).

Choose $\rho > 1$ such that $\rho\delta < d(K;E\setminus U)$ and let $s \in B(0;\delta)$. If $\Delta$ denotes the closed unit disc in $\mathbb{C}$, then $K + \Delta\rho s$ is a compact subset of $U$, and hence we can find $\varepsilon > 0$ such that the set $K + \Delta\rho s + B(0;\rho\varepsilon)$ is contained in $U$ and $f$ is bounded there, say by $M$. An application of the Cauchy integral formula implies that, for each $t \in B(0;\varepsilon)$,

$$|f^{(n)}(y)(s + t)| \leq \rho^{-n} \sup_{x \in K} |f^{(n)}(x)(\rho s + \rho t)| \leq \rho^{-n}M$$

Thus for each $s \in B(0;\delta)$ there exists $\varepsilon > 0$ such that the series $\Sigma_{n=0}^{\infty} f^{(n)}(y)(s + t)$ converges uniformly for $t \in B(0;\varepsilon)$. This shows that the series $\Sigma_{n=0}^{\infty} f^{(n)}(y)(s)$ defines a holomorphic function $f_1$ on the ball $B(y;\delta)$, and certainly $f_1 = f$, as we wanted.

The following theorem, which can be found in Noverraz [11], gives a necessary and sufficient condition for an open set to be a domain of existence.

2.7 THEOREM. Let $U$ be an open set in a separable Banach space $E$. Then $U$ is a domain of existence if and only if there is an increasing sequence of sets $A_j \subset U$ with the following properties:

(a) Each compact subset of $U$ is contained in some $A_j$.

(b) $d((\hat{A}_j)_{H(U)};E\setminus U) > 0$ for every $j$.

PROOF. First assume that $U$ is a domain of existence for a function $f \in H(U)$. For each $x \in U$ we choose $\varepsilon_x > 0$ such that $B(x;2\varepsilon_x) \subset U$ and $f$ is bounded on $B(x;2\varepsilon_x)$. Since $E$ is a separable metric space, the open cover $\{B(x;\varepsilon_x)|: x \in U\}$ of $U$ admits a countable subcover $\{B(x_j;\varepsilon_j)\}_{j=1}^{\infty}$, where $\varepsilon_j = \varepsilon_{x_j}$. For each $j$ let $A_j = \cup_{i=1}^{j} B(x_i;\varepsilon_i)$. Then clearly each compact subset of $U$ is contained in some $A_j$. We claim that $d((\hat{A}_j)_{H(U)};E\setminus U) \geq \delta_j$, where $\delta_j = \inf\{\varepsilon_1,\ldots,\varepsilon_j\}$. Indeed, if $x_0 \in (\hat{A}_j)_{H(U)}$ then it follows from the Cauchy inequalities that the series $\Sigma_{n=0}^{\infty} f^{(n)}(x_0)(s)$ defines a holomorphic function $f_1$ on $B(x_0;\delta_j)$ and $f_1 = f$ on a neighborhood of $x_0$. Since $U$ is a domain of existence for $f$ we conclude that $B(x_0;\delta_j) \subset U$.

Conversely, let us assume the existence of an increasing sequence of sets $A_j \subset U$ verifying (a) and (b). Without loss of generality we may assume that $(\hat{A}_j)_{H(U)} = A_j$ for every $j$. Let $D$ be a countable dense subset of $U$ and let $(x_j)$ be a sequence of points in $U$ with the property that each member of $D$ appears in the sequence $(x_j)$ infinitely many times. For each $x \in U$ let $B(x)$ denote the largest ball centered at $x$ and contained in $U$. Then $B(x) \not\subset A_j$ for every $x \in U$ and every $j$. After replacing $(A_j)$ by a subsequence, if necessary, we can find a sequence $(y_j)$ such that

$$y_j \in B(x_j)\backslash A_j \quad \text{and} \quad y_j \in A_{j+1}$$

for every $j$. Then we can find a sequence $(f_j) \subset H(U)$ such that

$$\sup_{A_j} |f_j| \leq 2^{-j}$$

and

$$|f_j(y_j)| \geq j + 1 + \left|\sum_{i=1}^{j-1} f_i(y_j)\right|$$

for every $j$. If we define $f = \Sigma_{j=1}^{\infty} f_j$ then $f \in H(U)$ and $|f(y_j)| \geq j$ for every $j$. Suppose that there are connected open sets $U_1$ and $U_2$ and a function $f_1 \in H(U_1)$ such that $U_1 \not\subset U$, $\emptyset \neq U_2 \subset U \cap U_1$, and $f_1 = f$ on $U_2$. Without loss of generality we may assume that $U_2$ is a connected component of $U \cap U_1$. Choose $a \in U_2$ and $b \in U_1\backslash U$ and let $C$ be a path entirely contained in $U_1$ and joining $a$ and $b$. Let $c$ be the first point in $C \cap \partial U_2$. Then $c \in U_1 \cap \partial U_2 \cap \partial U$. We will show that $f_1$ is not locally bounded at the point $c$, which is absurd. Take any $\varepsilon > 0$ such that $B(c;2\varepsilon) \subset U_1$ and choose $x \in D \cap U_2 \cap B(c;\varepsilon)$. Then $c \in B(x;\varepsilon) \cap \partial U$ and hence

$$B(x) \subset B(x;\varepsilon) \subset B(c;2\varepsilon) \subset U_1$$

Choose $j_1 < j_2 < j_3 < \cdots$ such that

$$x = x_{j_1} = x_{j_2} = x_{j_3} = \cdots$$

Then all points $y_{j_1}, y_{j_2}, y_{j_3}, \ldots$ lie in $B(x)$ and hence $f$ is unbounded on $B(x)$. But since $x \in U_2$ and $B(x) \subset U \cap U_1$ we conclude that $B(x) \subset U_2$. Since $f_1 = f$ on $U_2$ we conclude that $f_1$ is unbounded on $B(x) \subset B(c;2\varepsilon)$ and the proof is complete.

## 3. THE LEVI PROBLEM IN BANACH SPACES WITH A SCHAUDER BASIS

The following is the theorem of Gruman and Kiselman [3].

3.1 THEOREM. Every pseudoconvex open set in a Banach space with a Schauder basis is a domain of existence.

Before proving this theorem we need some preparation. Let E be a Banach space with a Schauder basis $(e_n)$. Let $E_n$ denote the vector subspace generated by $e_1, \ldots, e_n$, and let $\pi_n: E \to E_n$ denote the canonical projection. After passing to an equivalent norm on E we may (and will) assume that $\|\pi_n\| = 1$ for every n. Given an open subset U of E, we consider the sets

$$A_j(U) = \{x \in E : \|x\| \le j,\ d(x;E\backslash U) \ge \frac{1}{j} \text{ and } \|\pi_k(x) - x\| \le \frac{1}{3} d(x;E\backslash U) \text{ for all } k \ge j\}$$

$$B_j(U) = \{x \in E : \|x\| \le j,\ d(x;E\backslash U) \ge \frac{1}{2j} \text{ and } \|\pi_k(x) - x\| \le \frac{1}{2} d(x;E\backslash U) \text{ for all } k \ge j\}$$

These sets have the following properties:

(a) $A_j(U) \subset A_{j+1}(U)$, $B_j(U) \subset B_{j+1}(U)$, and $A_j(U) \subset B_j(U)$ for every j.

(b) Each compact subset of U is contained in some $A_j(U)$.

(c) $\pi_n(A_j(U)) \subset B_j(U)$ and $\pi_n(B_j(U)) \subset U$ whenever $n \ge j$.

(d) If U is pseudoconvex then

$$\widehat{(A_j(U))}_{Psc(U)} = A_j(U) \quad \text{and} \quad \widehat{(B_j(U))}_{Psc(U)} = B_j(U)$$

for every j.

We will only show that $\pi_n(A_j(U)) \subset B_j(U)$ whenever $n \ge j$. The other assertions are clear.

Let $x \in A_j(U)$ and let $n \ge j$. Then, on one hand:

$$d(\pi_n(x);E\backslash U) \ge d(x;E\backslash U) - \|\pi_n(x) - x\| \ge \frac{2}{3} d(x;E\backslash U) > \frac{1}{2j}$$

On the other hand, for each $k \ge j$:

$$\|\pi_k(\pi_n(x)) - \pi_n(x)\| = \|\pi_n(\pi_k(x) - x)\| \le \|\pi_k(x) - x\|$$
$$\le \frac{1}{3} d(x;E\backslash U) \le \frac{1}{2} d(\pi_n(x);E\backslash U)$$

3.2 LEMMA. Let E be a Banach space with a Schauder basis and let U be a pseudoconvex open set in E. Then given $f_n \in H(U \cap E_n)$ and $\varepsilon > 0$ there exists $f \in H(U)$ such that

$$\sup_{A_n(U)} |f - f_n \circ \pi_n| \leq \varepsilon$$

PROOF. Since $\pi_n(B_n(U)) \subset U$ we see that $f_n \circ \pi_n \in H(B_n(U))$, and in particular $f_n \circ \pi_n \in H(B_n(U) \cap E_{n+1})$. The set $B_n(U) \cap E_{n+1}$ is a compact subset of $U \cap E_{n+1}$ and

$$\widehat{(B_n(U) \cap E_{n+1})}_{Psc(U \cap E_{n+1})} = B_n(U) \cap E_{n+1}$$

An application of [5, Theorem 4.3.2] yields a function $f_{n+1} \in H(U \cap E_{n+1})$ such that

$$\sup_{B_n(U) \cap E_{n+1}} |f_{n+1} - f_n \circ \pi_n| \leq \frac{\varepsilon}{2}$$

By induction we can find a sequence $(f_j)_{j=n+1}^{\infty}$, with $f_j \in H(U \cap E_j)$, such that

$$\sup_{B_j(U) \cap E_{j+1}} |f_{j+1} - f_j \circ \pi_j| \leq \frac{\varepsilon}{2^{j-n+1}}$$

for every $j \geq n$. Hence

$$\sup_{A_j(U)} |f_{j+1} \circ \pi_{j+1} - f_j \circ \pi_j| = \sup_{\pi_{j+1}(A_j(U))} |f_{j+1} - f_j \circ \pi_j|$$

$$\leq \sup_{B_j(U) \cap E_{j+1}} |f_{j+1} - f_j \circ \pi_j| \leq \frac{\varepsilon}{2^{j-n+1}}$$

for every $j \geq n$. Thus for $k > j \geq n$ we obtain that

$$\sup_{A_j(U)} |f_k \circ \pi_k - f_j \circ \pi_j| \leq \sum_{i=j}^{k-1} \sup_{A_j(U)} |f_{i+1} \circ \pi_{i+1} - f_i \circ \pi_i|$$

$$\leq \sum_{i=j}^{\infty} \frac{\varepsilon}{2^{i-n+1}} = \frac{\varepsilon}{2^{j-n}}$$

Hence the sequence $(f_j \circ \pi_j)$ converges uniformly on each $A_i(U)$ to a function $f \in H(U)$. Furthermore, for each $j \geq n$:

$$\sup_{A_j(U)} |f - f_j \circ \pi_j| \leq \frac{\varepsilon}{2^{j-n}}$$

3.3 LEMMA. Let E be a Banach space with a Schauder basis and let U be a pseudoconvex open set in E. Then

$$\widehat{(A_j(U))}_{H(U)} \subset B_j(U)$$

for every j.

PROOF. Let $x \in \widehat{(A_j(U))}_{H(U)}$ and choose $k \geq j$ such that $x \in A_k(U)$. We will show that

$$\pi_n(x) \in \widehat{(B_j(U) \cap E_n)}_{H(U \cap E_n)} \tag{1}$$

for every $n \geq k$. Indeed, let $f_n \in H(U \cap E_n)$. Then given $\varepsilon > 0$, by Lemma 3.2 we can find $f \in H(U)$ such that

$$\sup_{A_n(U)} |f - f_n \circ \pi_n| \leq \varepsilon$$

Since $A_j(U) \cup \{x\} \subset A_k(U) \subset A_n(U)$ we obtain that

$$|f_n \circ \pi_n(x)| \leq |f(x)| + \varepsilon \leq \sup_{A_j(U)} |f| + \varepsilon$$

$$\leq \sup_{A_j(U)} |f_n \circ \pi_n| + 2\varepsilon \leq \sup_{B_j(U) \cap E_n} |f_n| + 2\varepsilon$$

Since $\varepsilon > 0$ was arbitrary, we conclude that

$$|f_n \circ \pi_n(x)| \leq \sup_{B_j(U) \cap E_n} |f_n|$$

and (1) is proven. Thus, by [5, Theorem 4.3.4] we obtain, for every $n \geq k$:

$$\pi_n(x) \in \widehat{(B_j(U) \cap E_n)}_{H(U \cap E_n)} = \widehat{(B_j(U) \cap E_n)}_{Psc(U \cap E_n)} = B_j(U) \cap E_n$$

Thus $\pi_n(x) \in B_j(U)$ for all $n \geq k$, and since $B_j(U)$ is closed, we conclude that $x \in B_j(U)$.

Theorem 3.1 follows at once from Lemma 3.3 and Theorem 2.7.

## 4. THE OKA-WEIL THEOREM IN BANACH SPACES WITH A SCHAUDER BASIS

The following theorem sharpens a result of Noverraz [12].

4.1 THEOREM. Let E be a Banach space with a Schauder basis, let U be a pseudoconvex open set in E and let K be a compact subset of U such that

$\hat{K}_{Psc(U)} = K$. Then for each $g \in H(K)$ there are an open set $V$ with $K \subset V \subset U$ and a sequence $(f_n) \subset H(U)$ such that $g \in H(V)$ and the sequence $(f_n)$ converges to $g$ uniformly on each compact subset of $V$.

Before proving this theorem we need some preparation.

4.2 DEFINITION. Let $U$ be an open set in a Banach space $E$.

(a) An open set $V \subset U$ is said to be $Psc(U)$-*convex* if $d(\hat{K}_{Psc(U)} \cap V; E\backslash V) > 0$ for each compact set $K \subset V$.

(b) An open set $V \subset U$ is said to be *strongly* $Psc(U)$-*convex* if $d(\hat{K}_{Psc(U)}; E\backslash V) > 0$ for each compact set $K \subset V$.

Observe that if $V$ is strongly $Psc(U)$-convex then $\hat{K}_{Psc(U)} \subset V$ for every compact set $K \subset V$.

4.3 EXAMPLES. Let $U$ be any open set in a Banach space. Then

(a) Every convex open set $V \subset U$ is strongly $Psc(U)$-convex.

(b) If $f \in Psc(U)$ then the open set $V = \{x \in U : f(x) < 0\}$ is strongly $Psc(U)$-convex.

(c) The intersection of finitely many (strongly) $Psc(U)$-convex open subsets of $U$ is (strongly) $Psc(U)$-convex.

4.4 THEOREM. Let $E$ be a Banach space with a Schauder basis, let $U$ be a pseudoconvex open set in $E$, and let $V$ be an open subset of $U$ which is strongly $Psc(U)$-convex. Then for each $g \in H(V)$ there exists a sequence $(f_n) \subset H(U)$ which converges to $g$ uniformly on each compact subset of $V$.

PROOF. For every $j$ and every $n$ the set $B_j(V) \cap E_n$ is a compact subset of $V \cap E_n$. Moreover,

$$\widehat{(B_j(V) \cap E_n)}_{Psc(U \cap E_n)} \subset \widehat{(B_j(V) \cap E_n)}_{Psc(U)} \cap E_n$$

Thus, since $V$ is strongly $Psc(U)$-convex, we may conclude that the set

$$\widehat{(B_j(V) \cap E_n)}_{Psc(U \cap E_n)}$$

is a compact subset of $V \cap E_n$. Let $g \in H(V)$ be given. Then in particular $g \in H(V \cap E_n)$ and an application of [5, Theorem 4.3.2] yields a function $h_n \in H(U \cap E_n)$ such that

$$\sup_{B_n(V) \cap E_n} |h_n - g| \le \frac{1}{n}$$

and hence

$$\sup_{A_n(V)} |h_n \circ \pi_n - g \circ \pi_n| \leq \frac{1}{n}$$

By Lemma 3.2, for each n we can find a function $f_n \in H(U)$ such that

$$\sup_{A_n(U)} |f_n - h_n \circ \pi_n| \leq \frac{1}{n}$$

We claim that $(f_n)$ converges to g uniformly on each compact subset of V. Indeed, let L be a compact subset of V and let $\varepsilon > 0$ be given. Choose $n_0 \geq 1/\varepsilon$ such that $L \subset A_{n_0}(V)$ and such that

$$\sup_L |g \circ \pi_n - g| \leq \varepsilon \qquad \text{for all } n \geq n_0$$

Then, for every $n \geq n_0$ we have that

$$\sup_L |f_n - g| \leq \sup_{A_n(V)} |f_n - h_n \circ \pi_n| + \sup_{A_n(V)} |h_n \circ \pi_n - g \circ \pi_n|$$

$$+ \sup_L |g \circ \pi_n - g| \leq \frac{1}{n} + \frac{1}{n} + \varepsilon \leq 3\varepsilon$$

Theorem 4.1 follows at once from Theorem 4.4 with the aid of Theorem 4.5 below, which is due to the author [9].

4.5 THEOREM. Let U be a pseudoconvex open set in a Banach space E, and let K be a compact subset of U such that $\hat{K}_{Psc(U)} = K$. Then, for each open set V with $K \subset V \subset U$ there is an open set W such that $K \subset W \subset V$ and W is strongly Psc(U)-convex.

PROOF. Choose $\varepsilon > 0$ such that $d(K;E\backslash U) > \varepsilon$. If we define

$$f(x) = -\log d(X;E\backslash U) + \log \varepsilon$$

then $f \in Psc(U)$ and

$$K \subset \{x \in U : f(x) < 0\}$$

Set

$$L = \bar{\Gamma}(K) \cap \{x \in U : f(x) \leq 0\}$$

where $\bar{\Gamma}(K)$ denotes the closed, convex hull of K. Since $\bar{\Gamma}(K)$ is a compact subset of E and since the set $\{x \in U : f(x) \leq 0\}$ is bounded away from $E\backslash U$, we conclude that L is a compact subset of U, and certainly $K \subset L$. Since

$\hat{K}_{Psc(U)} = K$, for each $a \in L\setminus V$ we can find a function $g_a \in Psc(U)$ such that $g_a(a) > 0$ and $g_a(x) < 0$ for every $x \in K$. By compactness of $L\setminus V$ we can then find functions $g_1, \ldots, g_m \in Psc(U)$ such that

$$g_j(x) < 0 \text{ for every } x \in K \text{ and every } j = 1, \ldots, m$$

and

$$L\setminus V \subset \bigcup_{j=1}^{n} \{x \in U : g_j(x) > 0\}$$

hence

$$L \cap \{x \in U : g_j(x) \leq 0 \text{ for } j = 1, \ldots, m\} \subset V$$

If we set $h = \sup\{f, g_1, \ldots, g_m\} \in Psc(U)$, then

(1) $h(x) < 0$ for every $x \in K$

(2) $\bar{\Gamma}(K) \cap \{x \in U : h(x) \leq 0\} \subset V$

We will show the existence of a positive integer n such that

(3) $(\bar{\Gamma}(K) + B(0;1/n)) \cap \{x \in U : h(x) \leq 0\} \subset V$

Otherwise we could find a sequence $(x_n)$ such that

$$x_n \in (\bar{\Gamma}(K) + B(0;1/n)) \cap \{x \in U : h(x) \leq 0\} \cap (E\setminus V)$$

For each n we choose $y_n \in \bar{\Gamma}(K)$ such that $\|x_n - y_n\| < 1/n$. Since $\bar{\Gamma}(K)$ is compact, the sequence $(y_n)$ admits a subsequence $(y_{n_k})$ which converges to a point $y \in \bar{\Gamma}(K)$. But then the corresponding subsequence $(x_{n_k})$ of $(x_n)$ converges to y too. But then

$$y \in \bar{\Gamma}(K) \cap \{x \in U : h(x) \leq 0\} \cap (E\setminus V)$$

contradicting (2). Thus (3) is proven and it suffices to set

$$W = (\bar{\Gamma}(K) + B(0;1/n)) \cap \{x \in U : h(x) < 0\}$$

## 5. FINAL REMARKS

A Banach space E is said to have the *approximation property* if given any compact set $K \subset E$ and any $\varepsilon > 0$, there is a finite rank continuous linear operator T on E such that $\|Tx - x\| \leq \varepsilon$ for every $x \in K$.

A Banach space E is said to have the *bounded approximation property* if there is a constant $M > 0$ such that, given any compact set $K \subset E$ and any $\varepsilon > 0$, there is a finite rank continuous linear operator T on E such that $\|T\| \leq M$ and $\|Tx - x\| \leq \varepsilon$ for every $x \in K$.

A Banach space with a Schauder basis clearly has the bounded approximation property. Conversely, Pelczynski has proved that every separable Banach space with the bounded approximation property is a complemented subspace of a Banach space with a Schauder basis (see [8, Theorem 1.e., 13]).

Using Pelczynski's result, it is not hard to extend Theorems 3.1, 4.1, and 4.4 to the case of separable Banach spaces with the bounded approximation property. In the case of Theorem 3.1, this was done by Noverraz [11].

A counterexample of Josefson [6] shows that Theorem 3.1 cannot be true for nonseparable Banach spaces.

It would be very important to determine the validity of Theorems 3.1, 4.1, and 4.4 for separable Banach spaces, or at least for separable Banach spaces with the approximation property.

## REFERENCES

1. H. Bremermann, Über die Äquivalenz der pseudokonvexen Gebiete und der Holomorphiegebiete in Raum von n Komplexen Veranderlichen, Math. Ann. 128 (1954), 63-91.

2. S. Dineen and A. Hirschowitz, Sur le théorème de Levi banachique, C. R. Acad. Sci. Paris 272 (1971), 1245-1247.

3. L. Gruman and C. O. Kiselman, Le problème de Levi dans les espaces de Banach à base, C. R. Acad. Sci. Paris 274 (1972), 1296-1298.

4. A. Hirschowitz, Prolongement analytique en dimension infinie, Ann. Inst. Fourier Grenoble 22 (1972), 255-292.

5. L. Hörmander, An Introduction to Complex Analysis in Several Variables, North-Holland, Amsterdam, 1973.

6. B. Josefson, A counterexample in the Levi problem, in Proceedings on Infinite Dimensional Holomorphy (T. Hayden and T. Suffridge, editors), Lecture Notes in Math. 364, Springer, Berlin, 1974, 168-177.

7. E. E. Levi, Sulle ipersuperfici dello spazio a 4 dimensioni che posono essere frontiera del campo di esistenza di una funzioni analitica di due variabili complesse, Ann. Mat. Pura Appl. 18 (1911), 69-79.

8. J. Lindenstrauss and L. Tzafriri, Classical Banach Spaces I, Springer, Berlin, 1977.

9. J. Mujica, Domains of holomorphy in (DFC)-spaces, in Functional Analysis, Holomorphy and Approximation Theory (S. Machado, editor), Lecture Notes in Math. 843, Springer, Berlin, 1981, 500-533.

10. F. Norguet, Sur les domains d'holomorphie des fonctions uniformes de plusieurs variables complexes, Bull. Soc. Math. France 82 (1954), 137-159.

11. Ph. Noverraz, Pseudo-Convexité, Convexité Polynomiale et Domaines d'Holomorphie en Dimension Infinie, North-Holland, Amsterdam, 1973.

12. Ph. Noverraz, Approximation of holomorphic or plurisubharmonic functions in certain Banach spaces, in Proceedings of Infinite Dimensional Holomorphy (T. Hayden and T. Suffridge, editors), Lecture Notes in Math. 364, Springer, Berlin, 1974, 178-185.

13. K. Oka, Sur les fonctions analytiques de plusieurs variables II, Domaines d'holomorphie, J. Sci. Hiroshima Univ. 7 (1937), 115-130.

14. K. Oka, Sur les fonctions analytiques de plusieurs variables VI, Domaines pseudoconvexes, Tohoku Math. J. 45 (1942), 15-52.

15. K. Oka, Sur les fonctions analytiques de plusieurs variables IX, Domaines finis sans point critique interieur, Japan, J. Math. 23 (1953), 97-155.

16. A. Weil, L'integral de Cauchy et les fonctions de plusieurs variables, Math. Ann. 111 (1935), 178-182.

# 13
# Calcul Infinitésimal en Géométrie Différentielle Synthétique

LUC BÉLAIR / Department of Mathematics, Yale University, New Haven, Connecticut

GONZALO E. REYES / Departement de Mathematiques et de Statistique, Université de Montreal, Montreal, Quebec, Canada

Nous présenterons ici quelques résultats reliés au calcul infinitésimal dans le contexte de la Géométrie différentielle synthétique (GDS). Le but de cette théorie est de donner une axiomatisation intrinsèque, directe ou naîve de la Géométrie différentielle afin de rendre explicite les raisonnements synthétiques employés par Darboux, Lie, Cartan et d'autres.

Kock (1981) décrit ce raisonnement comme étant celui qui "deals with space forms in terms of their structure, i.e., the basic geometric and conceptual constructions that can be performed on them".

Le tableau suivant essaie de présenter quelques traits caracteristiques des raisonnements synthétique et analytique:

---

Cette recherche a été subventionnée, en partie, par le Ministère de l'Education du Québec et le Conseil de recherches en sciences et en génie du Canada.

| synthétique | analytique |
|---|---|
| manipulation directe ("algébrique") des objets géométriques | manipulation analytique des représentations des objets géométriques |
| logique naîve | logique classique |
| arithmétique infinitésimale | limite |

Les sections §1, §2, §4 rappellent brièvement les notions de calcul infinitésimal utilisées. Dans les sections §5, §6, nous n'avons souvent qu'adapté les calculs classiques à notre contexte. Ces calculs se révèlent alors comme étant constructifs. On pourra trouver les détails de ces calculs ainsi que les preuves des theorèmes employés dans Bélair (1981).

En appendice on trouvera une preuve synthétique d'une version infinitésimale du théorème de Gauss-Bonnet et un dictionnaire définissant les notions de base du langage synthétique en termes des notions classiques des fonctions lisses et des idéaux de telles fonctions.

§.0

La droite réelle dont nour disposons est un anneau local ordonné $\langle R,+,\cdot,-,0,1,>\rangle$ i.e. un anneau unitaire commutatif où: si $x + y \# 0$ alors $x \# 0$ ou $y \# 0$ ($x \# 0 := x$ est inversible), $<$ est un ordre compatible avec la structure d'anneau, et $x \# 0$ ssi $x < 0$ ou $x < 0$.

La définition $x \leq 0 := \neg(x > 0)$ donne un pré-ordre compatible avec $<$ et la structure d'anneau, tel que $[0,0]$ contient tous les nilpotents.

Ces propriétés de $R$ suffisent à englober la majeure partie des situations ci-dessous bien qu'une fraction de celles-ci sied souvent à un espect donné. Par exemple, l'introduction du calcul différentiel ne requiert qu'un anneau unitaire commutatif, l'intégration que l'ajout d'un pré-order $\leq$ tel que $[0,0]$ contient les nilpotents.

## §1. LE CALCUL DIFFERENTIAL

On obtient une notion de dérivée en demandant que $R$ soit un objet de type ligne. $D$ désigne les éléments de carré nul i.e. $\{x \in R: x^2 = 0\}$, "les infiniments petits de 1[er] order".

1.1 Axiome de Kock-Lawvere (Dans l'infiniment petit toute courbe est une droite).

$$\forall f: D \to R \quad \exists! a,b \in R, \quad \forall d \in D \quad f(d) = a + bd \ .$$

Si $f: R \to R$, $x \in R$, alors $f(x+-): D \to R$, donc par l'axiome il existe $a,b$ uniques tels que $f(x + d) = a + bd$ pour $d \in D$. On voit que $a = f(x)$; on définit $f'(x) := b$ ainsi $f(x + d) = f(x) + df'(x)$.

Les règles habituelles de calcul sont valides: $(f+g)' = f'+g'$ etc. De façon analogue les dérivées partielles et directionnelles d'une fonction $f: R^m \to R^m$ sont définies; toute fonction est $C^\infty$.

Si $R$ est un anneau de Fermat au sens où pour $f: R \to R$ $F(x,y) = \dfrac{f(x) - f(y)}{x - y}$ possède une unique extension $\partial f: R^2 \to R$, on a tout une autre notion de dérivée: $f'(x) := \partial f(x,x)$. On peut alors formuler une règle de l'Hospital comme suit. Soient $f,g: R \to R$ tels que $f(0) = 0 = g(0)$, $\partial g(x,0) \# 0$ pour tout $x$, alors il existe un unique $h: R \to R$ t.g. $h(x) = \dfrac{f(x)}{g(x)}$ si $x \# 0$ et $h(0) = \dfrac{f'(0)}{g'(0)}$ .

on montre que la règle de l'Hospital est vérifiée si et seulement si pour tout $F: R \to R$, si $F(0) = 0$ et $F(x) = 0$ pour tout $x \# 0$, alors $F$ est identiquement nulle. Dans notre contexte, cette condition et l'axiome d'intégration (c.f. §2) assurent que $R$ est un anneau de Fermat.

## §2. LE CALCUL INTEGRAL

Les primitives permettent d'intégrer.

**2.1 Axiome d'intégration.** (Toute fonction admet une primitive)

$$\forall f: [0,1] \to R, \quad \exists! g: [0,1] \to R, \quad g' = f \wedge g(0) = 0$$

On définit alors $\int_0^x f(t)\, dt := g(x)$ où $g$ est donné par l'axiome et les bonnes propriétés de $\int_0^1$ s'ensuivent immédiatement.

**2.2 Proposition.**

(1) $$\int_0^1 a\, f(t)\, dt = a\int_0^1 f(t)\, dt\,, \qquad a \in R$$

(2) $$\int_0^1 (f+g)(t)dt = \int_0^1 f(t)dt + \int_0^1 g(t)dt$$

(3) $$f(b) - f(a) = (b-a)\int_0^1 f'(a + (b-a)t)dt$$

i.e. le lemme d'Hadamard

(4) $$\frac{d}{ds}\int_0^1 f(s,t)dt = \int_0^1 \frac{\partial f}{\partial s}(s,t)dt$$

(5) $$\int_0^1 f'(t)g(t)dt = f(1)g(1) - f(0)g(0) - \int_0^1 f(t)g'(t)dt$$

démonstration: par l'unicité dans 2.1: on considère les deux membres de l'égalité comme fonctions sur $[0,1]$ et on vérifie qu'elles ont même dérivée et coïncident en $0$. □

2.3 Proposition. ($R$ est connexe).

Soit $f: [a,b] \to R$. Si $f' = 0$ sûr $[a,b]$ alors $f$ est constante.

L'intégration est bien définie sur tous les intervalles de $R$: l'axiome d'intégration sur $[0,1]$ donne une propiété analogue sûr tous les intervalles (bien sûr les propriétés de $<$ et $\leq$ jouent ici beaucoup). Ces extensions de $\int_0^1$, toutes compatibles entre elles, héritent de ses propriétés.
Par exemple, $\int_a^b f(x)dx = (b-a)\int_0^1 f(a+(b-a)t)dt$

2.4 Proposition.

(1) Soit $a \leq b \leq c$ $\int_a^b f(t)dt + \int_b^a f(t)dt = \int_a^c f(t)dt$

(2) Soit $a_i \leq b_i$, $\int_{a_1}^{b_1}\left(\int_{a_2}^{b_2} f(x,y)dy\right)dx = \int_{a_2}^{b_2}\left(\int_{a_1}^{b_1} f(x,y)dx\right)dy$

(3) Soit $g: [c,d] \to R$, $f: [a,b] \to R$ t.g. $f(a) = c$

$$\int_a^{f(t)} g(x)dx = \int_0^t g \circ f(x)\ f'(x)dx$$

(4) $$f(x) = \sum_{k=0}^{n} \frac{f^{(k)}(a)}{k!}(x-a)^k + \int_a^x f^{(n+1)}(t)\frac{(x-t)^n}{n!}dt$$

(5) $$f(x) = \sum_{k=0}^{n} \frac{f^{(k)}(0)}{k!}x^k$$

$$+\ x^{n+1}\int_0^1 \cdots \int_0^1 f^{(n+1)}(xu_1\ldots u_{n+1})u_n\ldots u_1^n du_1\ldots du_n$$

Le comportement numérique de l'intégrale prend un sens grâce à l'axiome suivant.

2.5 Axiome de positivité de l'intégrale.

$$\forall f: [0,1] \to R \quad ((\forall x \in [0,1] \quad f(x) > 0) \to \int_0^1 f(t)dt > 0)$$

Ce dernier s'étend bien à tout intervalle $[a,b]$ par le lemme d'Hadamard. La positivité de $\int$ sur les intervalles ouverts n'est pas immédiate. En effet on a recours à $(f(a)f(b) < 0) \to \neg(\forall t \in [a,b] \; f(t) \# 0)$ qui est valide en supposant R pythagoricien, ou encore par l'axiome dit de la valeur intermédiaire transversale (cf. §.3)

2.6 Axiome. (R est pythagoricien)

$$\forall x \in R \quad (x > 0 \quad \exists y \in R \quad y > 0 \wedge y^2 = x).$$

2.7 Proposition.

Soit $f: [a,b] \to R$, $a < b$. Si $f(a)f(b) < 0$ alors $\neg(\forall t \in [a,b] \quad f(t) \# 0)$.

démonstration: On considère $F(x) = \dfrac{f(x)}{\surd(f^2(x))}$ et on utilise la connexité au sens de 2.3 □

2.8 Corollaire.

Si $f > 0$ sur $(a,b)$ alors $\int_a^b f(t)dt > 0$.

## §.3 LES FONCTIONS TRIGONOMETRIQUES

L'existence d'une solution au système d'équations différentielles classique suffit à donner l'unicité et les propriétés élémentaires des solutions.

3.1 Axiome d'existence des fonctions trigonométriques.

$$\exists u,v: R \to R, \quad u' = v \wedge v' = -u \wedge u(0) = 0 \wedge v(0) = 1.$$

3.2 Proposition.

(1) Pour $u,v$ comme dans l'axiome, $u^2 + v^2 = 1$

(2) Le couple $(u,v)$ de l'axiome est unique. On note

$\sin := u, \quad \cos := v.$

(3) $\sin(x + y) = \sin x \cos y + \cos x \sin y$

$\cos(x + y) = \cos x \cos y - \sin x \sin y.$

(4) cos est une fonction paire, sin impaire.

démonstration: les preuves habituelles. □

Pour introduire $\pi/2$, le premier zéro positif de cos, on a recours à une propriété analogue au théorème de la valeur intermédiaire. Celle ci en diffère toutefois un peu car il n'est pas vérifié dans plusieurs modèles.

3.3 Axiome de la valeur intermediaire transversale.

$$\forall f: [0,1] \to R((f(0)f(1) < 0 \wedge \forall x \in (0,1) \quad (f(x) = 0 \to f'(x) \# 0))$$
$$\to \exists c \in (0,1) \; (f(c) = 0 \wedge \forall x \in (0,c) \quad f(0)f(x) > 0)$$

L'axiome dit que si une fonction change de signe sur un intervalle et ne peut couper l'axe que transversalement, alors elle le coupe une première fois.

3.4 Proposition.

(1) Unicité de la valeur intermédiaire transversale.

(2) $\cos 2 < 0$.

De (3.4) decoule l'existence de $\pi/2$ comme premier zéro positif de cos au sens de l'axiome (3.3).

3.5 Proposition.

(1) $\sin \pi/2 = 1$

(2) $2\pi$ est une periode de sin et cos

(3) $\cos x > 0$ pour $-\pi/2 < x < \pi/2$

(4) $1 < \pi/2 < 2$.

Pour illustrer certain type d'argument utilisé ici voyons qu'étant donné $\cos x > 0$ pour tout $0 < x < 1$, alors $1 \leq \pi/2$ i.e. $\neg(1 > \pi/2)$. Supposons $1 > \frac{\pi}{2}$, on pose $f(t) = -t + \frac{\pi}{2}$, alors $f(0)f(1) < 0$ et $f'(t) = -1 \# 0$. Il existe donc $0 < t_o < 1$ t.g. $t_o = \frac{\pi}{2}$ par l'axiome de la valeur intermédiaire transversale, ce qui contredit notre première hypothèse $\cos t_o > 0$.

3.6 Proposition.

(1) $\sin\colon (\frac{-\pi}{2}, \frac{\pi}{2}) \to (-1,1)$ est un isomorphisme, l'inverse étant

$$\arcsin t = \int_0^t \frac{1}{\sqrt{(1 - x^2)}}\, dx.$$

(2) $\mathrm{cis}\colon R \to S^1$, $\mathrm{cis}(\theta) = (\cos \theta, \sin \theta)$, est surjectif. En fait $\mathrm{cis}\colon (-\pi, \frac{3\pi}{2}) \to S^1$ l'est.

(3) $2\pi$ est la plus petite période de cos au sens où si $T$ est une période de cos et $\frac{-5\pi}{2} < T < \frac{5\pi}{2}$ alors $T = 0$ ou $T = 2\pi$ ou $T = -2$ .

(4) $f\colon (\frac{-3\pi}{2}, \frac{3\pi}{2}) \to R$ se factorise par $\mathrm{cis}\colon (\frac{-3\pi}{2}, \frac{3\pi}{2}) \to S^1$ si et seulement si $f(x + 2\pi) = f(x)$ pour tout $x \in (\frac{-3\pi}{2}, \frac{3\pi}{2})$.

## §.4 FORMES ET INTEGRATION

On ne rappellera ici que les définitions et les résultats clés. Le lecteur est invité à consulter Kock-Reyes-Veit (1980) ou Kock (1981).

Les formes sont des fonctions de poids sur les voisinages infinitésimaux des points d'un objet. De tels voisinages sont donnés par des applications $\gamma: D^n \to M$. On définit une intégration sur les cubes infinitésimaux $c \in M^{D^n} \times D^n$ qu'on prolonge ensuite à des cubes finis $c \in M^{I^n}$, $I = [0,1]$, par "sommation" de toutes les intégrations infinitésimales.

### 4.1 Définition.

(1) Les n-cubes infinitésimaux sur $M := M^{D^n} \times D^n$

(2) Les n-chaînes infinitésimales sur $M := C_n(M)$, la notion habituelle de R-combinaison linéaire formelle de n-cubes infinitésimaux.

(3) Les n-cochaînes sur $M := C^n(M)$, les $\omega : M^{D^n} \to R$ avec $(a\omega)(c) = a.\omega(c)$, $a \in R$.

(4) $\int : C_n(M) \times C^n(M) \to R$, définie par linéarité sur les n-cubes infinitésimaux $<c,\underline{d}>$, $\underline{d} = (d_1,\dots,d_n)$: $\int_{<c,\underline{d}>} \omega := d_1 \dots d_n \omega(c)$

(5) Un n-cube infinitésimal $<c,\underline{d}>$ est dégénéré si $d_i = 0$ pour un $i$, $1 \le i \le m$.

### 4.2 Proposition.

Il y a correspondence biunivoque entre les applications $\Omega: C_n(M) \to R$ qui s'annulent sur les n-cubes infinitésimaux dégénérés et les n-cochaînes $\omega: M^{D^n} \to R$.

**4.3 Definition.** $\partial: C_{n+1}(M) \to C_n(M)$ est définie par linéarité.

Pour $\gamma = \langle c, \underline{d} \rangle$,

$$\partial\gamma = \sum_{i=1}^{n} \sum_{\alpha=0,1} (-1)^{i+\alpha} F_{i\alpha}(\gamma)$$

où $F_{i\alpha}(\gamma) = \langle c_i(\alpha d_i), \underline{d}(\hat{\imath}) \rangle$ et $\underline{d}(\hat{\imath})$ est le n-tuplet obtenu de $\underline{d}$ en enlevant la i-ème composante, e.g. pour $n = 1$, $i = 1$, $c_1(\delta): D \to M$, $c_1(\delta)(d) = c_1(\delta, d)$, $\underline{d}(\hat{1}) = d_2$.

Pour une n-cochaîne fixeé $\omega$, on a une fonction $\int_{\partial -} \omega: M^{D^{n+1}} \times D^{n+1} \to R$ qui s'annule sur les n+1-cubes infinitésimaux dégénérés d'où l'existence d'une n+1-cochaîne $\Omega_\omega$ entièrement déterminée par $\omega$ et telle que $\int_{\partial\gamma} \omega = \int_\gamma \Omega_\omega$.

**4.4 Définition.** Dans la discussion précédente on définit $d\omega := \Omega_\omega$. Il s'ensuit qu'on a un théorème de Stokes infinitésimal par définition même de $d\omega$, la dérivée extérieure de $\omega$.

**4.5 Proposition.**

(1) $\int$ est "fonctoriel" i.e. $\int_\gamma f^\star\omega = \int_{f^\star\gamma} \omega$.

(2) $\partial \circ \partial = 0$, $d \circ d = 0$

(3) $\partial, d$ sont linéaires et fonctoriels.

**4.6 Definition.** Une forme différentielle sur $M$ est une cochaîne que possède les propriétés

(1) alternance: $\omega(c \circ D^\pi) = \text{sign}(\pi)\omega(c)$ où $\pi$ est une permutation, $c \circ D^\pi (d_1, \ldots, d_n) = c(d_{\pi(1)}, \ldots, d_{\pi(n)})$.

(2) homogénéité: $\omega(a_k \cdot c) = a\omega(c)$ où $1 \le k \le n$, $a_k \cdot c(d_1, \ldots, d_n) = c(d_1, \ldots, ad_k, \ldots, d_n)$.

La dérivée exérieure est fermée sur les formes. Toute n-forme sur $R^m$ peut s'écrire de la façon canonique habituelle. Par exemple pour les 2-formes sur $R^2$, $\omega = f(x,y)\, dx \wedge dy$ ou $f: R^2 \to R$ et pour $c: D^2 \to R^2$

$$c(d_1,d_2) = \underline{p} + \underline{x}_1 d_1 + \underline{x}_2 d_2 + \underline{x}_{12} d_1 d_2$$

$$\underline{p} = (p_1,p_2), \quad \underline{x}_i = (x_{i1},x_{i2})$$

$$e: D \to R^2, \quad e(d) = (x+ud, y+vd)$$

on a $$dx, dy: R^{2^D} \to R, \quad dx(c) = u, \quad dy(e) = v$$

$$dx \wedge dy: R^{2^D} \to R, \quad dx \wedge dy(c) = x_{11}x_{22} - x_{12}x_{21} = \det(\underline{x}_1,\underline{x}_2)$$

Intégration des formes sur les chaînes finies

On a les notions analogues de n-cube $M$, $c \in M^{I^n}$, $I = [0,1]$ de n-chaîne, $CI_n(M)$; et de bord $\partial$. Si c'est un $\ell$-cube et $c_2$ un m-cube alors $c_1 \times c_2$ est un $\ell$+m-cube et on étend ce produit à $CI_n(M)$. On a les 1-cubes $[[a,b]]: I \to R$, $[[a,b]](t) = a + (b-a)t$, $a \leq b \in R$, et donc les n-cubes $[[a_1,b_1]] \times \ldots \times [[a_n,b_n]]$ appelés n-rectangles. On définit une relation $\sim$ sur les chaînes telle que $[[a,b]] + [[b,c]] \sim [[a,c]]$, $a \leq b \leq c$ de façon à pouvoir ensuite définir une intégrale des formes sur les chaînes finies, compatible avec cette relation.

4.7 Proposition. Si $\gamma_1 \sim \gamma_2$ alors $\partial\gamma_1 \sim \partial\gamma_2$.

4.8 Définition. Un n-rectangle $[[a_1,b_1]] \times \ldots \times [[a_n,b_n]]$ est D-petit si $b_i - a_i \in D$ pour tout $1 \leq i \leq m$; dégénéré si $b_i - a_i = 0$ pour un $1 \leq i \leq m$.

4.9 Théorème. Soit $\Phi$ une fonction à valeurs dans $R$ définie sur les n-rectangles D-petits, qui s'annule sur les n-rectangles dégénérés. Il existe une extension additive unique $\overline{\Phi}$ de $\Phi$ aux n-rectangles, additive au sens où si $\rho \sim \rho_1 + \rho_2$ alors $\overline{\Phi}(\rho) = \overline{\Phi}(\rho_1) + \overline{\Phi}(\rho_2)$.

C'est ce théorème qui permet d'intégrer les formes sur les chaînes finies. En effet si c est un n-cube, $\omega$ une n-forme, alors $c^*\omega$ est une n-forme sur $I^n$; on peut donc, par image réciproque, se ramener au cas des n-rectangles. Or nous possèdons déjà une intégration sur les n-rectangles D-petits puisqu'on peut les assimiler à des n-cubes infinitésimaux. Pour c et $\omega$ donnés cette intégration est une fonction sur les rectangles D-petits satisfaisant les hypothèses du théorème: $\int_c \omega$ sera l'image du rectangle unité $([[0,1]] \times \ldots \times [[0,1]])$ par $\overline{\Phi}$, en fait

$$\int_c \omega = \int_0^1 \ldots \int_0^1 \omega([<\delta_1,\ldots,\delta_m> \mapsto c(x_1+\delta_1,\ldots,x_m+\delta_m)])dx_1\ldots dx_m$$

4.10 Proposition. Soit $c_i$, c, $\partial$ des chaînes.

(1) Si $c_1 \sim c_1 + c_2$ alors $\int_c \omega = \int_{c_1} \omega + \int_{c_2} \omega$ .

(2) $\int_{f_*\gamma} \omega = \int_\gamma f^*\omega$

(3) $\int_{\partial c} \omega = \int_c d\omega$

démonstration: (3) il s'agit de vérifier que l'intégration sur les chaînes est compatible avec l'intégration sur les chaînes infinitésimales, le résultat découle du théorème de Stokes infinitésimal. □

4.11 Proposition.

(1) $\mathrm{cis}^D: (-\pi,\frac{3\pi}{2})^D \to (S^1)^D$ $\mathrm{cis}^D(c) = \mathrm{cis} \circ c$, est surjectif.

(2) Soit $g: R \to M$ un isomorphisme, $f: M \to R$ une o-forme.

Si $df = 0$ alors f est constante.

On dira que e.g. $c: [a,b] \to M$ est un 1-cube puisqu'on peut considérer $I \xrightarrow{A} [a,b] \xrightarrow{c} M$, A l'application affine évidente.

## §.5 LES GROUPES DE COHOMOLOGIE DE DE RHAM

La cohomologie de de Rham est définie de la façon habituelle: soit $\mathcal{M}^n(M)$ := les n-formes sur $M$, $F^n(M) = \{\omega \in \mathcal{M}^n(M)^S : d\omega = 0\}$, les n-formes fermées, $E^n(M) = \{\omega \in \mathcal{M}^n(M) : \exists\, \alpha \in \mathcal{M}^{n-1}(M) \quad d\alpha = \omega\}$, les n-formes exactes. Ce sont tous des R-modules, et $E^n(M) \subseteq F^n(M)$. Le n-ième groupe de cohomologie de de Rham de $M$ est le R-module quotient $H^n(M) := F^n(M)/E^n(M)$. $H^M$ est fonctoriel de façon contravariante.

5.1 Proposition.

(1) $H^n(R) = 0$, $n = 1,2.,,,$

(2) $H^2(R^2) = 0$.

démonstration: la remarque qu'on puisse écrire les formes sur $R^n$ de la façon cononique habituelle permet d'utiliser les arguments standards.

La calculs subséquents sont adaptés de ceux de Flanders dans "Studies in global geometry and analysis"; le premier cas donnera une idée de l'adaptation des preuves.

5.2 Proposition. $H^1(S^1) = R$, $S^1 = \{(x,y) \in R^2 : x^2 + y^2 = 1\}$.

démonstration: Soit le 1-cube $cis: [0,2\pi] \to S^1$, $cis(t) = (\cos t, \sin t)$, $\partial cis = cis(2\pi) - cis(0) = 0$. Considérant $\int_{cis} - : F^1(S^1) \to R$, on a $E^1(S^1) \subset \ker \int_{cis} \_$ donc $\exists!\ f: H^1(S^1) \to R$ t.g. $f \circ v = \int_{cis} \_$ où $v$ est la projection canonique. On montre que $f$ est un isomorphisme.

(i) $f$ *est surjectif*: soit $\delta\theta$ la forme d'angle sur $S^1$, i.e. $\delta\theta = -y\,dx + x\,dy$. Alors $\int_{cis} \delta\theta = 2\pi \neq 0$, et $2\pi$ engendre $R$.

(ii) f *est injectif*: si $\int_{cis} \omega = 0$, pour $\omega$ fermée, alors est exacte. Soit $\omega \in F^1(S^1)$ tel que $\int_{cis} \omega = 0$

$$U^+ = \{(x,y) \in S^1 : y > -\tfrac{1}{2}\}, \quad U^- = \{(x,y) \in S^1 : y < \tfrac{1}{2}\}.$$

Alors $S^1 = U^+ \cup U^-$, $U^+ U^-$ isomorphes à R. Puisque $\mathcal{M}^1(R) = E^1(R)$ on a

$\omega = d\alpha$ sur $U^+$, pour quelque $\alpha: U^+ \to R$

$\omega = d\beta$ sur $U^-$, pour quelque $\alpha: U^- \to R$

Soit $H^+ = cis\restriction_{[0,\pi]}$, $H^- = cis\restriction_{[\pi,2\pi]}$, alors on a

$$\int_{H^+} \omega = \int_{H^+} d\alpha = \int_{H^+} \alpha = \alpha(-1,0) - \alpha(1,0)$$

de même, $\int_{H^-} \omega = \beta(1,0) - \beta(-1,0)$

aussi $0 = \int_{cis} \omega = \int_{H^+} \omega + \int_{H^-} \omega$ car $[0,2\pi] \sim [0,\pi] + [\pi,2\pi]$ donc

$\alpha(-1,0) - \beta(-1,0) = \alpha(1,0) - \beta(1,0) = k \in R$.

Sur $U^+ \cap U^-$, $d(\alpha - \beta) = d\alpha - d\beta = 0$ et $U^+ \cap U^- = A \cup B$, A,B disjoints isomorphes à R.

$$\Rightarrow \alpha = \beta + k.$$

Posons $\gamma: S^1 \to R$, $\gamma(z) = \begin{cases} \alpha(z) & \text{si } z \in U^+ \\ \beta(z) + k & \text{si } z \in U^- . \end{cases}$

Comme $(S^1)^D = U^{+D} \cup U^{-D}$, $d\gamma = \omega$ puisqu'ils coincident sur un recouvrement de $(S^1)^D$. □

5.3 Proposition.

(1) $H^1(B^2) = 0$, $B^2 = \{(x,y) \in R^2 : x^2 + y^2 \leq 1\}$.

(2) $H^1(R^2) = 0$.

Ces calculs donnent le théorème de point fixe de Brouwer. Pour $x \in R^n$, $x \# 0 := x.x \# 0$, le produit scalaire habituel.

**5.4 Corollaire.** Si $f: B^2 \to B^2$, alors $\neg \forall x \quad B^2 \in f(x) \# x$.

démonstration: l'argument habituel; on utilise le fait suivant:
$a,b \in R^2$, $b.b \# 0$, $a.a \leq 1$, $(a-b) \leq 1$ alors
$(a.b)^2 + b.b(1-a.a) > 0$. □

**5.5 Proposition.**

(1) $H^1(C) = R$, $C = S^1 \times (-1,1)$ le cylindre

(2) $H^2(S^2) = R$, $S^2 = \{(x,y,z) \in R^3: x^2 + y^2 + z^2 = 1\}$,

(3) $H^1(S^2) = 0$

(4) $H^1(\underline{P}^2) = 0$ $\underline{P}^2$ le plan projectif.

démonstration: (2) On doit postuler l'existence d'une fonction plate
AXIOME: $\exists F:R \to R$, $\forall x \in R \quad (x < 0 \to F(x) = 0 \wedge x > 0 \to F(x) > 0)$.

(4) $\underline{P}^2$ est le coegalisateur de $1_{S^2}$ et de l'application antipode. □

## §.6 LE DEGRE DE BROUWER D'UNE FONCTION

Pour $f: S^1 \to S^1$, le calcul $H^1(S^1) = R$ permet de définer le degré de $f: H^1(f)(\delta\theta) = H^1(f)(\delta\theta + E^1(S^1)) = \lambda\delta\theta + E^1(S^1)$ pour un unique $\lambda \in R$, par définition $\deg f := \lambda$. En fait $\deg f = \frac{1}{2\pi}\int_{cis} f^*\delta\theta$.

**6.1 Proposition.** Soit $a,b, \theta_o \in R$, $f: [a,b] \to S^1$ tel que $cis\ \theta_o = f(a)$. Il existe une unique fonction $F: [a,b] \to R$ $cis \circ F = f$ et $F(a) = \theta_o$.

démonstration: $F(t) = \int_0^t (y'(s)x(s) - x'(s)y(s)ds + \theta_o$, où
$f(s) = (y(s),x(s))$. □

**6.2 Corollaire.** (deg f est entier) Pour $f: S^1 \to S^1$, $2\pi \deg f$ est une période de cis cad $\mathrm{cis}(2\pi \deg f) = (1,0)$.

démonstration: soit $F$ tel que $\mathrm{cis} \circ F = f \circ c$ où $c = \mathrm{cis} \restriction [0,2\pi]$.

$$\begin{aligned} 2\pi \deg f &= \int_c f^* \delta\theta = \int_{c_*[[0,2\pi]]} f^*\delta\theta \\ &= \int_{[[0,2\pi]]} c^* f^* \delta\theta = \int_{[[0,2\pi]]} F^* \mathrm{cis}^* \delta\theta \\ &= \int_{[[0,2\pi]]} F^* dt = \int_{[[0,2\pi]]} dF \\ &= F(2\pi) - F(0) \end{aligned}$$

et $\mathrm{cis}(F(2\pi)) = f(c(2\pi)) = f(c(0)) = \mathrm{cis}(F(0))$. □

**6.3 Corollaire.** (Invariance homotopique du degré)

Soit $F: [0,1] \times S^1 \to S^1$, $\deg(F(0,-)) = \deg(F(1,-))$.

démonstration: Posons $g(t) = \deg(F(t,-1): [0,1] \to R$.

$g(t + d) = g(t) + dg'(t)$ pour $d \in D$.

$\Rightarrow$ $\deg(F(t+d,-1) - \deg(F(t,-1) = dg(t)$ est une période de cis

$\Rightarrow$ $\sin(dg'(t)) = dg'(t) = 0$, pour tour $d \in D$

$\Rightarrow$ $g'(t) = 0$

$\Rightarrow$ $g$ est constant. □

**6.4 Proposition.**

(1) Si

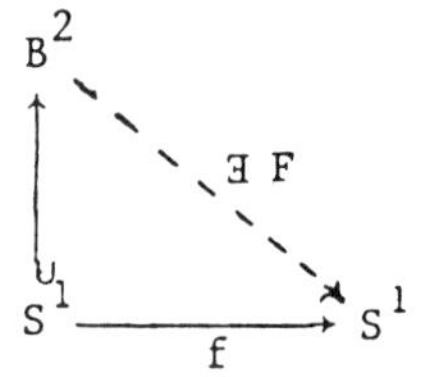

$$\begin{array}{ccc} B^2 & & \\ \uparrow u_1 & \searrow \exists F & \\ S^1 & \xrightarrow{f} & S^1 \end{array}$$

commute alors $\deg f = 0$.

(2) Si $h: S^1 \to S^1$ préserve les antipodes alors

$\text{cis}(\pi \deg h) = (-1,0)$ i.e. "deg h est impair".

De 6.4 on tire les théorèmes de Gauss-D'Alembert et de Borsuk-Ulam. Soit $C = R[i]$, $i^2 + 1 = 0$, on écrit $z \# 0$ pour dire que $z$ est inversible.

6.5 Proposition.

(1) Soit $p(Z) = Z^n + c_{n-1}Z^{n-1} + \ldots + c_o$, $n > 1$ un polynôme complexe. Alors $\neg \forall z \in C \; p(z) \# 0$.

(2) Si $f: S^2 \to R^2$ alors $\neg \forall x \in S^2 \; f(x) \# f(-x)$.

## §.7 LES COURBES PLANES ET LE THEOREME D'EULER

On développe ici une théorie des courbes planes et on considère ensuite le théorème d'Euler sur la courbure normale des courbes sur une surface.

7.1 Définition. Une courbe $\gamma$ est une application $\gamma: [0,1] \to R^2$ t.g. $\gamma'(t) \# 0$ pour tout $t \in [0,1]$.

Pour une courbe $\gamma$, on a l'application "vecteur tangent normalisé" $t_\gamma: [0,1] \to S'$, $t_\gamma(s) = \frac{\gamma'(s)}{\|\gamma'(s)\|}$. La courbuse est donnée par le d'enroulement des recteurs tangents sur le cercle.

7.2 Définition. Soit $\gamma$ une courbe, il existe $F: [0,1] \to R$ tel que $\text{cis} \circ F = t_\gamma$, on définit la courbure $K_\gamma(s) = \frac{F'(s)}{\|\gamma'(s)\|}$

7.3 Proposition.

(1) Soit $\gamma = (\gamma_1, \gamma_2)$ une courbe $K = \frac{\gamma_1'\gamma_2'' - \gamma_1''\gamma_2'}{(\gamma_1'^2 + \gamma_2'^2)^{3/2}}$

(2) Soit $K: [0,1] \to R$, il existe une courbe $\gamma$ t.g. $K = K_\gamma$

(3) Soit $\gamma_1, \gamma_2$ deux courbes t.g. $K_{\gamma_1} = K_{\gamma_2}$, $\|\gamma_1'\| = 1 = \|\gamma_2'\|$, il existe un unique mouvement euclidien qui envoie $\gamma_1$ sur $\gamma_2$.

Ce qui suit est relié au théorème d'Euler. Comme les infinitésimaux de $1^{er}$ order, D, avaient décrit la tangence, les infinitésimaux de $2^{er}$ ordre, $D_2 = \{x \in R: x^3 = 0\}$, décriront la courbure.

7.4 Axiome. (Toute fonction sur $D_2$ est analytique)

$$\forall f: D_2 \to R, \quad \exists!\ a,b,c \in R, \quad \forall u \in D_2 \quad f(\ ) = a + ub + u^2c.$$

Cet axiome nous permet de définir $f'(u) = b + 2uc$, $f''(u) = 2c$.

7.5 Corollaire

(1) Pour $f: R \to R$, $u \in D_2$, $f(x+u) = f(x) + uf'(x) + u^2 \frac{f''(x)}{2}$

(2) Pour $F,G: D_2 \to R$, $(FG)'(0) = F'(0)G(0) + F(0)G'(0)$

$(F \circ G)'(0) = F'(G(0))G'(0)$.

7.6 Définition. Un arc infinitesimal lié au point $p \in R^2$ est une application $c: D_2 \to R^2$ telle que $c(0) = p$, $c'(u) \# 0$ pour $u \in D_2$.

On a une notion analogue de courbure pour les arcs infinitésimaux, qui est compatible avec la première notion.

7.7 Définition. Soit $f: E \to R$, $E \subseteq R$ tel que $c+d \in E$ $\forall e \in E$ $\forall d \in D$. $f(x)$ est un maximum (resp. minimum) de $f$ si $f'(x) = 0$ et si pour tout $y \in E$ $f(y) \leq f(x)$ (resp. $f(y) \geq f(x)$).

7.8 Proposition.

(1) Soit $S \subseteq R^3$, donné par $f(x,y,z) = 0$, $p \in S$ tel que $f_x(p) \# 0$ ou $f_y(p) \# 0$ ou $f_z(p) \# 0$. Il existe un changement de coordonnées qui envoie $S$ sur $F(X,Y,Z) = 0$ et $p$ à l'origine, de façon à ce que $F_X(0) = 0$, $F_Y(0) = 0$, $F_Z(0) \# 0$.

(2) (Théorème des fonctions implicites infinitésimal). Soit $F: R^3 \to R$ comme en (1), alors $\forall_{\varepsilon,\delta \in R}(\varepsilon^3 = \delta^3 = \varepsilon^2\delta = \varepsilon\delta^2 = 0$ $\to \exists !\, \eta \in D \quad F(\varepsilon,\delta,\eta) = 0)$.

(3) Soit $S \subseteq R^3$ associé à $F(X,Y,Z) = 0$, comme en (1). Si il existent exactement deux valeurs extrêmes $\lambda_1 \# \lambda_2$ de la courbure des courbes obtenues par l'intersection de $S$ avec des plans parallèles à la normale à la surface en $0$ et de direction $\text{cis}\ \theta$ alors ces extrema sont atteints dans exactement deux directions perpendiculaire $\text{cis}\ \theta_1$, $\text{cis}\ \theta_2$ au sens où $\text{cis}\ \theta_1 \cdot \text{cis}\ \theta_2 = 0$.

(4) Soit $S$ et $F(X,Y,Z)$ comme en (3), alors $\lambda_1$, $\lambda_2$ existent si et seulement si $F_{XX}(0) \# F_{YY}(0)$ ou $F_{XY}(0) \# 0$.

Soit $S$, $F(X,Y,Z)$ comme précédemment, $K$ désigne la courbure normale sur $S$ en $0$ i.e. la courbure des courbes obtenues par l'intersection de $S$ avec des plans contenant la normale en $0$.

7.9 Théorème d' Euler. Suppons que $R$ soit un corps au sens de Kock i.e. $\neg(x_1 = 0 \wedge x_2 = 0 \wedge \ldots \wedge x_n = 0) \to (x_1 \# 0 \vee x_2 \# 0 \vee \ldots \vee x_m \# 0)$. Si $K$ n'est pas constante $(\neg K = \text{constante})$ alors il existe exactement deux extrema pour $K$ dans exactement deux directions perpendiculaires.

démonstration: $K(\theta) = A \cos^2\theta + B \sin^2\theta + 2C \sin\theta \cos\theta$

$$A = \frac{-F_{XX}(0)}{F_Z(0)}, \qquad B = \frac{-F_{YY}(0)}{F_Z(0)}, \qquad C = \frac{-F_{XY}(0)}{F_Z(0)}$$

et on ulitise 7.8. □

## APPENDICE 1

### Version infinitesimale de Gauss-Bonnet

On supposera que $M$ a une *structure infinitésimale conforme 2-dimensionnelle* au sens où chaque fibre $\pi^{-1}(x) \subset M^D$ est un R-module libre de rang 2 et qu'on dispose d'une mesure pour les angles infiniment petits entre deux vecteurs non nuls de $\pi^{-1}(x)$ telle que $\measuredangle(\gamma,\gamma) = 0$ et $\measuredangle(\gamma_1,\gamma_2) + \measuredangle(\gamma_2,\gamma_3) = \measuredangle(\gamma_1,\gamma_3)$. On notera que $\measuredangle(\gamma_2,\gamma_1) = -\measuredangle(\gamma_1,\gamma_2)$.

Une donnée de *transport parallèle* sur M est une fonction qui associe à chaque $h \in D$ et chaque courbe $\gamma \in M^D$ une bijection

$$\tau_h(\gamma,-): \pi^{-1}(\gamma(0)) \xrightarrow{\sim} \pi^{-1}(\gamma(h))$$

telle que

$$\begin{cases} \tau_o(\gamma,\delta) = \delta \\ \tau_h(a\gamma,\delta) = \tau_{ah}(\gamma,\delta)\ ,\ \tau_h(\gamma,a\delta) = a\tau_h(\gamma,\delta) \\ \tau_h \text{ preserve les angles .} \end{cases}$$

Un exemple d'une telle structure est "un espace de Riemann ordonné de dimension 2 avec le transport parallèle canonique".

On peut donner une explication heuristique de la *courbure* d'une surface en termes de l'angle que fait un vecteur tangent non nul avec le vecteur obtenu par transport parallèle le long d'un cycle "très petit":

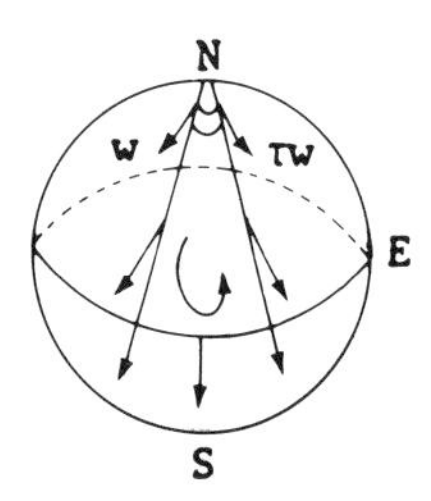


Dans notre contexte tout en suivant E. Cartan (Leçons sur la géométrie des espaces de Riemann), cette explication permet de définir la *forme de courbure* du transport parallèle en identifiant un tel cycle avec un parallélogramme infinitésimal $(\gamma,(h_1,h_2))$

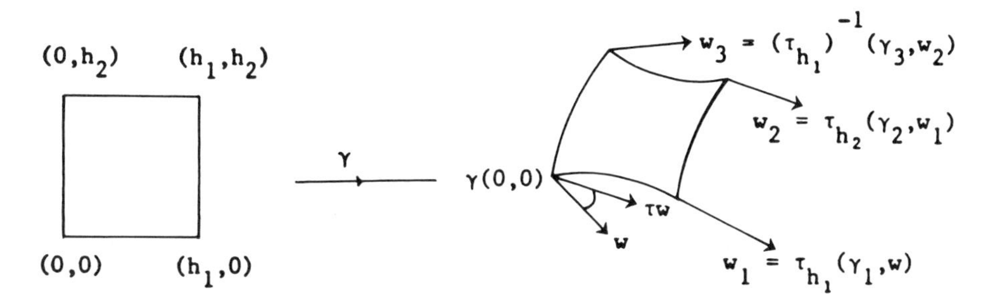


Ici, $\tau w = (\tau_{h_2})^{-1}(\gamma_4, w_3)$, $\gamma_1(h) = \gamma(h,0)$, $\gamma_2(h) = \gamma(h_1,h)$, etc...

On remarque que $\measuredangle(\tau w, w)$ s'annule quand $h_1 = 0$ et aussi quand $h_2 = 0$. D'autre part, l'axiome de Kock-Lawvere montre que $\measuredangle(\tau w, w)$ est de la forme $A_o + A_1 h_1 + A_2 h_2 + A_{12} h_1 h_2$. Ces deux remarques impliquent que $A_o = A_1 = A_2 = 0$. Le nombre $A_{12}$ dépend seulement de $\gamma$ et on le note $K(\gamma)$, la *forme de courbure*. On notera que $K(\gamma)$ ne dépend pas de $w$, car le transport parallèle préserve les angles:

$$\measuredangle(\tau w', w') = \measuredangle(\tau w', \tau w) + \measuredangle(\tau w, w) + \measuredangle(w, w') = \measuredangle(\tau w, w).$$

En regardant la figure précédente on ne peut pas s'empêcher de voir une intégration le long du bord et la question se pose: existe-t-il une 1-forme $\phi$ telle que $d\phi = K$? Le théorème de Gauss-Bonnet répond à cette question.

Dans une région ouverte $U \subset M$ pour laquelle il existe un champ de vecteurs non-nuls $X: U \to U^D$, on considère

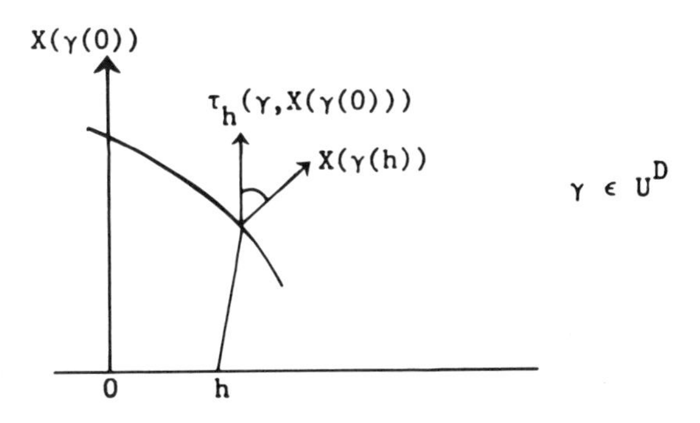


$\gamma \in U^D$

et on définit la *forme de connexion* $\phi_X$ par la formule

$$\sphericalangle(\tau_h(\gamma,X(\gamma(0))),X(\gamma(h))) = -h\phi_X(\gamma) = -\int_{(\gamma,h)} \phi_X .$$

Theoreme (Gauss-Bonnet Infinitésimal)

$$d\phi_X = -K$$

En particulier, $d\phi_X$ ne dépend pas du choix du champ de vecteurs non-nuls $X$.

Preuve. Par observation de la figure suivante

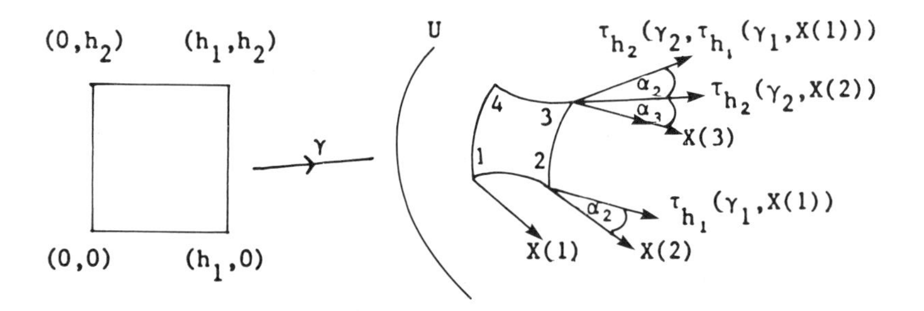


Comme tout à h'heure $\gamma_1(h) = \gamma(h,0)$, $\gamma_2(h) = \gamma(h_1,h)$, $\alpha_4 = \sphericalangle((\tau_h)^{-1}(\gamma_3,X(3)),X(4))$, etc... On a

$$-\int_{\partial(\gamma,(h_1,h_2))} \phi_X = -\int_{(\gamma_1,h_1)} \phi_X - \int_{(\gamma_2,h_2)} \phi_X + \int_{(\gamma_3,h_1)} \phi_X + \int_{(\gamma_4,h_2)} \phi_X$$

$$= \alpha_2 + \alpha_3 + \alpha_4 + \alpha_1 .$$

D'autre part, en transportant $X(1)$ le long du cycle 1 2 3 4 1 on "gagne" les degrés suivants par rapport au champ $X$: $\alpha_2$ au point 2, $\alpha_2 + \alpha_3$ au point 3, $\alpha_2 + \alpha_3 + \alpha_4$ au point 4 et $\alpha_2 + \alpha_3 + \alpha_4 + \alpha_1$ au point 1. Autrement dit

$$\sphericalangle(\tau X(1),X(1)) = \alpha_2 + \alpha_3 + \alpha_4 + \alpha_1 = h_1 \cdot h_2 \cdot K(\gamma) = \int_{(\gamma,(h_1,h_2))} K .$$

APPENDICE 2

## Dictionnaire d'interprétation

On donne ici une interprétation des notions de base de la théorie synthétique GDS en termes des notions classiques des fonctions lisses et des idéaux de telles fonctions.

Soit $I \subset C^\infty(\mathbb{R}^n)$ un idéal (au sens algébrique) de la $\mathbb{R}$-algèbre des fonctions lisses (cad, possédant des dérivés partielles continues) de $\mathbb{R}^n$ dans $\mathbb{R}$.

(i) *Un réel au stade* $(n,I)$ est une classe d'équivalence $f \bmod I$, où $f \in C^\infty(\mathbb{R}^n)$. Les réels sont donc des "réels variables" dépendant lissement d'un paramètre.

(2) $0, 1, +, ., -$ *au stade* $(n,I)$:

$$(f \bmod I) + (g \bmod I) = (f + g) \bmod I$$
$$(f \bmod I) . (g \bmod I) = (f . g) \bmod I$$
$$-(f \bmod I) = -f \bmod I$$
$$1 = 1 \bmod I$$
$$0 = 0 \bmod I$$

(3) *Un réel dans* D *au stade* $(n,I)$ est une classe $f \bmod I$, où $f^2 \in I$.

(4) *Un réel* $> 0$ au stade $(n,I)$ est une classe $f \bmod I$ telle que $\exists g \in C^\infty(\mathbb{R}^n)$ $\chi(f(x)).g(x) = 1 \bmod I$ où $\chi \in C^\infty(\mathbb{R})$ est une fonction telle que $\chi(x) \neq 0$ ssi $x > 0$.

(5) *Un reel dans* $[0,1]$ *au stade* $(n,I)$ est une classe $f \bmod I$ telle que $\forall \phi \ (\phi|_{[0,1]} \equiv 0 \to \phi\ (f(x)) \in I)$

(6) *Une fonction dans* $R^R$ *au stade* $(n.I)$ est donneé par $F \bmod I$ , ou $F \in C^\infty(\mathbb{R}^n \times \mathbb{R})$ et $I^*$ est l'idéal dans

$C^\infty(\mathbb{R}^n \times \mathbb{R})$ engendré par $\{f \circ \pi \mid f \in I\}$, $\pi: \mathbb{R}^n \times \mathbb{R} \to \mathbb{R}$. Les fonctions sont, elles aussi, des "fonctions variables de t" $F(x,t)$ dépendant lissement du paramètre $x$.

(7) *Une fonction dans* $R^D$ *au stade* $(n.I)$ est une classe $F \bmod (I.t^2)$ où $F \in C^\infty(\mathbb{R}^n \times \mathbb{R})$.

(8) *Une fonction dans* $R^{[0,1]}$ *au stade* $(n,I)$ est une classe $F \bmod (I,\{\phi \mid \phi \mid_{[0,1]} \equiv 0\})$.

(9) *L'évaluation d'une fonction à un point au stade* $(n,I)$ est donnée par $F(x,f(x)) \bmod I$, où $F(x,t) \bmod I$ est la fonction et $f(x) \bmod I$ le réel en question.

On remarquera que ce dictionnaire est fonctoriel au sens suivant: on organise les stades dans une *catégorie*, en définissant une flèche $(n,I) \xrightarrow{\phi} (m,J)$ comme étant une fonction $\phi \in C^\infty(\mathbb{R}^m, \mathbb{R}^n)$ telle que $f \circ \phi \circ J$ pour tout $f \in I$. Une fonction dans $R^R$ au stade $(n,I)$ définit alors, par composition avec $\phi \times Id$, une fonction dans $R^R$ au stade $(m,J)$. La même chose est valide pour les autres notions.

Pour compléter notre dictionnaire, il faudrait interpréter aussi les *connecteurs logiques* de la logique naive en termes des connecteurs de la logique classique. En effet, la logique classique est incompatible avec l'axiome de Kock-Lawvere (voir e.g. Kock (1981)). On y arrive en employant le "forcing de faisceaux" (voir e.g. Bélair (1981) ou Kock (1981) pour les détails).

On se contentera ici de vérifier l'axiome de Kock-Lawvere: une fonction $f \in R^D$ au stade $(n,I)$ est, selon 7), donnée par $F(x,t) \bmod (I,t^2)$. En employant le lemme d'Hadamard deux fois, on obtient $G \in C^\infty(\mathbb{R}^n \times \mathbb{R})$ telle que

$$F(x,t) = F(x,0) + t.\frac{\partial F}{\partial t}(x,0) + t^2 G(x,t), \text{ i.e.}$$

$$F(x,t) = F(x,0) + t.\frac{\partial F}{\partial t}(x,0) \bmod (I,t^2)$$

Cad, $f(d) = a + bd \quad \forall\, d \in D.$

L'axiome d'intégration présente des difficultés "analytiques", mais on y arrive en employant le résultat suivant: soit $M_X^\infty$ l'idéal de fonctions plates sur X (cad s'annulant, ainsi que leur dérivées, sur X).

Théorème. Si $X \subseteq \mathbb{R}^n$ et $Y \subseteq \mathbb{R}^m$ sont des fermés, alors $M_{X\times Y}^\infty = (M_X^\infty, M_Y^\infty)$.

(Voir Quê-Reyes (1982) pour la preuve et Reyes (1982) pour l'approche "toposophique" qui explique et étend ce dictionnaire).

REFERENCES


[1] L. Bélair, Calcul Infinitésimal en Géometrie différentielle synthétique, Thèse de maîtrise, Université de Montréal, 1981.

[2] E. Cartan, Leçons sur la Géometrie des espaces de Riemann, Gauthier-Villars, 1951.

[3] S.S. Chern (Ed.), Studies in Global Geometry and Analysis, *MAA Studies in Mathematics*, Vol.4, 1967.

[4] M.P. do Carmo, Differential Geometry of Curves and Surfaces, Prentice-Hall, 1976.

[5] E. Dubuc, $C^\infty$-schemes, *American Journal of Mathematics*, Vol.103, No.4, 1981.

[6] A. Kock, Synthetic Differential Geometry, *London Math. Soc. Lecture Notes Series 51,* Cambridge University Press, 1981.

[7] A. Kock, G.E. Reyes, B. Veit, Forms and integration in Synthetic Differential Geometry, *Aarhus Preprint Series,* 1979/1980, No.31, 1980.

[8] Ngo van Quê, G.E. Reyes, Smooth functors and synthetic calculus, à paraître dans *Brouwer's Centenary Proceedings.*

[9] G.E. Reyes, Modèles de la Géometrie différentielle synthétique, à paraître.

14

# Relating Topological Entropy to Finite Dynamical Systems

SERVET A. MARTINEZ / Departamento de Mathemáticas y Ciencias de la Computación, Facultad de Ciencias Físicas y Matemáticas, Universidad de Chile, Santiago, Chile

## 1. INTRODUCTION

By a (compact) dynamical system [d.s.] we mean a couple (X,T), X being a metric and compact space and T: X→X continuous.

We say that some property *P* is (topological) conjugate invariant if it is satisfied by the equivalence class of conjugate systems:

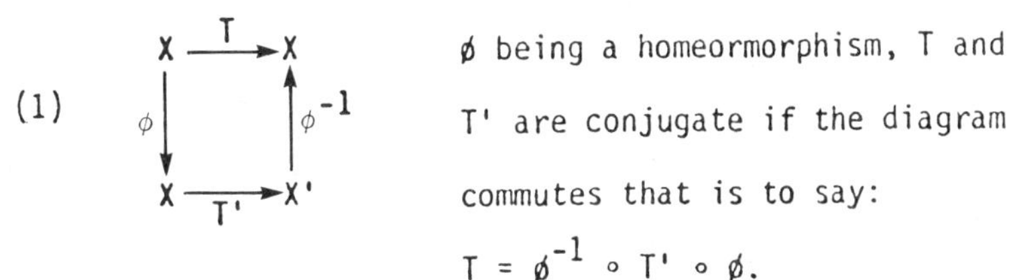


$\phi$ being a homeormorphism, T and T' are conjugate if the diagram commutes that is to say: $T = \phi^{-1} \circ T' \circ \phi$.

We will define the topological entropy, it is easy to prove that it is a conjugate invariant.

If $\mathcal{U} = (U_\alpha)_{\alpha\in A}$ is an open cover of X we will note $T^{-n}\mathcal{U} = (T^{-n}U_\alpha)_{\alpha\in A}$, which is also an open cover. If $\mathcal{U} = (U_\alpha)_{\alpha\in A}$ and $\mathcal{U}' = (U')_{\beta\in B}$ are two open covers we write $\mathcal{U}\vee\mathcal{U}' = (U_\alpha \cap U'_\beta)_{(\alpha,\beta)\in A\times B}$. Also we will note $(\mathcal{U})_0^n = \mathcal{U}\vee T^{-1}\mathcal{U}\vee \ldots \vee T^{-n}\mathcal{U}$.

If we note by $N(\mathcal{U})$ the smallest cardinality of a subcover of $\mathcal{U}$ we have that the function $S_n = \log(N(\mathcal{U})_0^{n-1})$ is positive and subadditive. Then the following limit exists:

$H(\mathcal{U},T) = \lim_{n\to\infty} \frac{1}{n} \log(N(\mathcal{U})_0^{n-1})$, it will be called the topological entropy of T with respect to $\mathcal{U}$.

The following number (may be infinite):
$H(T) = \text{Sup } \{H(\mathcal{U},T) \ / \ \mathcal{U}$ open cover of $X\}$ will be called the topological entropy of T. This is the definition given in [1]. In reference [3] many properties of this invariant are given.

In [2] Bowen has introduced the following definitions and results (that remain valid for non-compact dynamical systems).

Let us call $E \subset X$ $(n,\epsilon)$-separated if for any two points $x,y \in E$ there exists $0 \leq k < n$ such that $d(T^k x, T^k y) > \epsilon$. Denote by $s_n(\epsilon,T)$ the largest cardinality of an $(r,\epsilon)$ - separated subset of X. Also call $F \subset X$ $(n,\epsilon)$ - spanning, if for every $x \in X$ $\exists\, y \in F$ such that $d(T^k x, T^k y) \leq \epsilon$ for $0 \leq k < n$. Let us note by $r_n(\epsilon,T)$ the minimal cardinality of a $(n,\epsilon)$ - spanning set. By the uniform continuity of T we have that $s_n(\epsilon,T)$ and $r_n(\epsilon,T)$ are finite for every $\epsilon > 0$, $n \in \mathbb{N}$.

It is not hard to see that $r_n(\epsilon,T) \leq s_n(\epsilon,T) \leq r_n(\epsilon/2, T)$. These inequalities and the fact that $r_n(\epsilon,T)$ and $s_n(\epsilon,T)$ increases when $\epsilon$ decreases prove that the following quantities exist and they satisfy the equality:

$$\lim_{\epsilon\to 0} \limsup_{n\to\infty} \frac{1}{n} \log s_n(\epsilon,T) = \lim_{\epsilon\to 0} \limsup_{n\to\infty} \frac{1}{n} \log r_n(\epsilon,T)$$

Using these equality, Bowen proved that they are equal to the topological entropy (see [3] page 84) so we have the formula:

$$h(T) = \lim_{\epsilon\to 0} \limsup_{n\to\infty} \frac{1}{n} \log r_n(\epsilon,T) \tag{2}$$

## 2. CONJUGACY AND FINITE DYNAMICAL SYSTEMS

By a finite d.s. we will mean a couple $(Y,S)$ where $Y$ is a finite set and $S$ is a function $S:Y\to Y$.

Obviousy a non-finite d.s. cannot be conjugate (1) to a finite d.s. We will say that a d.s. $(X,T)$ is $(\epsilon,n)$ - conjugate to a finite d.s. $(Y,S,)$ if there exist $\phi,\psi$ so that:

a) $\phi: Y \to X$ satisfy $\psi\circ\phi = id_Y$

$\psi: X \to Y$

b) The following diagram commutes up to $\epsilon$ for the iterates 0 to n-1:

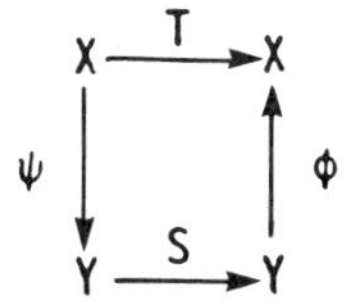

$$\begin{array}{ccc} X & \xrightarrow{T} & X \\ \psi\downarrow & & \uparrow\phi \\ Y & \xrightarrow{S} & Y \end{array}$$

$$\tilde{d}(T^k,(\phi S\psi)^k)\leq \epsilon \qquad 0\leq k<n;$$

where $$\tilde{d}(T,T') = \sup\{d(Tx, T'x) \ / \ x\in X\}.$$

From a) we obtain that $\phi$ is injective and $\psi$ is surjective. From a) and b) it is deduced that: $\tilde{d}(T^k,\phi S^k\psi) \leq \epsilon \quad 0\leq k\leq n$.

We will note by $\ell_n(\epsilon,T)$ the minimal cardinality of a set $Y$ which supports a finite d.s. $(\epsilon,n)$-conjugate to a given d.s. $(X,T)$.

First of all we shall proved that $\ell_n(\epsilon,T)$ is finite and we will relates it to $r_n(\epsilon,T)$.

LEMMA 1. $\ell_n(\epsilon,T) \leq n\ r_n(\epsilon,T)$

PROOF: Let $E\subset X$ be an $(n,\epsilon)$ - spanning set of $(X,T)$ such that card $E = r_n(\epsilon,T)$. Take $Y = \bigcup_{k=0}^{n-1} T^k E$, define over $Y$:

$S(T^k e) = T^{k+1}e$ if $e \in E$, $0 \leq k < n-1$

$S(T^{n-1}e) = e'$ if $e' \in E$ satisfied:

$d(T^k(T^n e), T^k e') \leq \epsilon$ for $0 \leq k < n$.

Then we have: $d(T^{k+n}e, T^k S T^{n-1}e) \leq \epsilon$ when $0 \leq k \leq n$.

Define $\psi(T^k e) = T^k e$ $0 \leq k < n$:

If $x \notin Y$ define $\psi(x) = T^{k_0}e$ with $0 \leq k_0 < n$, $e \in E$ anyone that satisfies $d(T^k x, T^{k+k_0}e) \leq \epsilon$ for $0 \leq k < n$ (the existence is by the $(n,\epsilon)$ - spanning property of E, so we can take $k_0 = 0$).

Let $\phi$ be the injection of Y in X then we have $\psi \circ \phi = id_Y$

If $x \notin Y$ we have:

$d(T^k x, S^k \psi(x)) \leq \epsilon$ by definition of $\psi$.

If $x = T^{k_0}e \in Y$, that is if $0 \leq k_0 < n$ we have:

$d(T^k x, S^k \psi(x)) = d(T^{k+k_0}e, S^k(T^{k_0}e))$ which is $\leq \epsilon$ by definition of S if $k+k_0 < n$. If $n \leq k+k_0 \leq 2(n-1)$ this last expression is equal to: $d(T^{\ell-1}(T^n e), S^{\ell-1}S(T^{n-1}e))$ for some $1 \leq \ell \leq n-1$. But $S(T^{n-1}e) = e' \in E$ where $S^{\ell-1}(e') = T^{\ell-1}(e')$ so we obtain:

$$d(T^k x, S^k \psi(x)) = d(T^{\ell-1}(T^n e), T^{\ell-1}S(T^{n-1}e)) \leq \epsilon \quad \text{for } 1 \leq \ell \leq n.$$

Then the result.

Q.E.D.

We have also a lower bound for $\ell_n(\epsilon,T)$:

<u>LEMMA 2.</u> $r_n(2\epsilon,T) \leq \ell_n(\epsilon,T)$

<u>PROOF</u>: Let (Y,S) be a finite d.s. which is $(n,\epsilon)$ - conjugate to (X,T) and card $Y = \ell_n(\epsilon,T)$.

The subset $E = \phi(Y)$ is an $(n,2\epsilon)$- spanning set because:

$$d(T^k x, T^k \circ \phi \circ \psi(x)) \leq d(T^k x, \phi \circ S^k \circ \psi(x)) + d(\phi \circ S^k \circ \psi(x)), T^k \circ \phi \circ \psi(x))$$

and $\phi \circ S^k \circ \psi \circ \phi \circ \psi = \phi \circ S^k \circ \psi$, so the last quantity is $\leq$ than $2\epsilon$.

Q.E.D.

PROPOSITION: $h(T) = \lim_{\epsilon \to 0} \limsup_{n \to \infty} \frac{1}{n} \log \ell_n(\epsilon, T)$

PROOF: From the Bowen's equality (2) and Lemmas 1, 2.

Q.E.D.

REFERENCES

[1] ADLER, R., A. KONHEIM, M. Mc ANDREW. "Topological Entropy" Trans. Amer. Math. Soc. 114(1965), 309-319.

[2] BOWEN, R. "Entropy for group endomorphisms and homogenous spaces" Trans. Amer. Math. Soc. 153(1971), 401-413.

[3] DENKER, M., C. GRILLENBERG, K. SIGMUND. "Lectures notes in Mathematics", Berlin-Heidelberg-New York, Springer 527(1976).

# 15
# Deformation Affine d'un Revêtement Ramifié

VICTOR GONZALEZ-AGUILERA / Departamento de Matemáticas, Universidad Técnica Federico Santa María, Valparaíso, Chile

## INTRODUCTION

Soit U un polydisque de $\mathbb{C}^n$, et $\ell\cdot X$ un revêtement ramifié (avec multiplicitée $\ell$) de U contenu dans U × $\mathbb{C}^p$. Le but de ce travail est de définir une deformation canonique "dite affine" du revêtement $\ell\cdot X$.

Pour cela on introduit tout d'abord une sous variété algèbrique d'un produit symétrique, cette variété qui en général n'est pas normal on l'appellera "l'accroché", et on la dénotera par $\mathrm{Sym}^k(\mathbb{C}^p) \not\swarrow \mathrm{Sym}^\ell(\mathbb{C}^q)$.

Elle nous permet de classifier les rêvetements ramifiés de degré $\ell\cdot k$ de U contenus dans U × $\mathbb{C}^p$ × $\mathbb{C}^q$ avec image directe donnée $\ell\cdot X$ contenu dans U × $\mathbb{C}^p$.

Finalement la donnée d'une application analytique de U dans l'accroché Sym $(\mathbb{C}^p) \not\swarrow \mathrm{Sym}^\ell(\mathbb{C}^q)$ nous permet définir d'une manière naturelle une deformation canonique de $\ell\cdot X$, qu'on appelle deformation affine.

## §0. RAPPELS ET NOTATIONS

$S(\mathbb{C}^p)$ dénotera l'algèbre symétrique sur $\mathbb{C}^p$, c'est à dire l'algèbre des polynômes sur $(\mathbb{C}^p)^*$ le dual de $\mathbb{C}^p$, on identifiera $S_h(\mathbb{C}^p)$ avec le $\mathbb{C}$-espace vectoriel des polynômes homogênes de degré h, ainsi $\overset{k}{\underset{1}{\oplus}} S_h(\mathbb{C}^p)$ est un $\mathbb{C}$-espace vectoriel de dimension $\binom{p+k}{k}-1$, qu'on note $E_p$.

$\mathrm{Sym}^k(\mathbb{C}^p)$ est la variété algèbrique normale quotient de $(\mathbb{C}^p)^k$ par l'action du groupe symétrique $S_k$, elle est plongée canoniquement dans l'espace $E_p$ à travers les fonctions symétriques tensorielles.

On appellera n-ième fonction symétrique de Newton de l'élement $(s_1, \ldots, s_k)$ de $E_p$, l'élément de $S_n(\mathbb{C}^p)$ défini par recurrence par les relations suivantes:

$$\sum_{h=0}^{n} (-1)^{n-h} N_{n-h} \cdot S_h = (k-n) S_n \qquad n \geqslant 0$$

avec les conventions $S_0 = 1$ et $S_n = 0$ si $n > k$.

On appellera automorphisme de Newton de $E_p$ l'isomorphisme algèbrique de $E_p$ dans $E_p$ définit par $s \to \overset{k}{\underset{1}{\oplus}} N_n(s)$. On dira qu'une application d'un ensemble X dans $E_p$ est définie en Newton si nous connaisons cette application f par les composantes homogènes de la application $(\overset{k}{\underset{1}{\oplus}} N_n) \circ f$.

## §1. REVÊTEMENTS RAMIFIÉS AVEC IMAGE DIRECTE DONNÉE

Définition 1

Soit U un polydisque de $\mathbb{C}^n$ et B un polydisque de $\mathbb{C}^p$ (qui peut être $\mathbb{C}^p$). Nous appellerons revêtement ramifié de degré k de U contenu dans U × B, la donnée d'un nombre fini de sous-ensembles analytiques irréductibles $X_i$ fermés dans $U \times B$, affectés des multiplicités $n_i > 0$, tel que la restriction à chaque $X_i$ de la projection naturelle $U \times B \to U$, soit propre surjective et de degré $k_i$, de sorte que l'on ait

$$\sum_i n_i \cdot k_i = k$$

On appellera support de X, noté $|X|$, le revêtement ramifié de U contenu dans $U \times \mathbb{C}^p$ défini par $X = (X_i, 1)\ i \in I$.

Soit Y un revêtement ramifié de degré $k\ell$ de U contenu dans $U \times \mathbb{C}^p \times \mathbb{C}^q$, on sait d'aprês (G. §1 Def. 1) définir l'image directe de Y par la projection partielle $U \times \mathbb{C}^p \times \mathbb{C}^q \to U \times \mathbb{C}^p$.

Dans cette section on se propose de clasifier tous les revêtements ramifiés de degré $k \cdot \ell$ de U contenus dans $U \times \mathbb{C}^p \times \mathbb{C}^q$ tels que leur image directe est $\ell \cdot X$ où X est un revêtement ramifié de degré k (sans multiplicités) de U contenu dans $U \times \mathbb{C}^p$, pour cela on va introduire une sous variété algèbrique de $\mathrm{Sym}^{k\ell}(\mathbb{C}^p \times \mathbb{C}^q)$ qu'on appellera l'accroche.

## Construction de $\mathrm{Sym}^k(\mathbb{C}^p) \checkmark \mathrm{Sym}^\ell(\mathbb{C}^q)$

Soit $S_{k\ell}$ le groupe de permutations de $k\ell$ lettres, on dénotera par H le sous groupe de $S_{k\ell}$ engendré par les permutations du type $\rho \circ \sigma$, où $\sigma \in \prod_1^k S_\ell$ (produit direct) et $\rho$ est une permutation de $S_{k\ell}$ qui permute chaque block de $\ell$ éléments en les voyants comme un seul élement.

On dénotera par $\mathrm{Sym}_k^\ell\ (\mathbb{C}^p \times \mathbb{C}^q)$ la variété algébrique normale quotient de $(\mathbb{C}^p \times \mathbb{C}^q)^{\ell k}$ par l'action de H (H est un groupe fini d'automorphismes).

Dans $(\mathbb{C}^p \times \mathbb{C}^q)^{\ell k}$ on considère les coordonnées suivantes $(x_1^1, y_1^1), \ldots \ldots, (x_\ell^1, y_\ell^1), \ldots \quad \ldots, (x_1^k, y_1^k), \ldots \ldots, (x_\ell^k, y_\ell^k)$.

La sous variété algèbrique V de $(\mathbb{C}^p \times \mathbb{C}^q)^{\ell k}$ définitie par les equations $x_1^1 = x_2^1 \ldots \quad \ldots = x_\ell^1, \ldots \quad \ldots x_1^k = x_2^k \ldots . = x_\ell^k$ est invariante par l'action de H, alors le quotient V/H s'identifie (comme on est en caracteristique 0) à une sous variété algèbrique de $\mathrm{Sym}_k^\ell(\mathbb{C}^p \times \mathbb{C}^q)$ en fait elle s'identifie à $\mathrm{Sym}^k(\mathbb{C}^p \times \mathrm{Sym}^\ell(\mathbb{C}^q))$.

Proposition 1

L'identité id : $(\mathbb{C}^p \times \mathbb{C}^q)^{\ell k} \to (\mathbb{C}^p \times \mathbb{C}^q)^{\ell k}$ induit une application algèbrique finie :

$$m : \mathrm{Sym}^k(\mathbb{C}^p \times \mathrm{Sym}^\ell(\mathbb{C}^q)) \to \mathrm{Sym}^{\ell k}(\mathbb{C}^p \times \mathbb{C}^q).$$

Démonstration

Comme H est un sous groupe de $S_{\ell k}$ on a une application algèbrique finie: id: $\mathrm{Sym}_k^\ell(\mathbb{C}^p \times \mathbb{C}^q) \to \mathrm{Sym}^{\ell k}(\mathbb{C}^p \times \mathbb{C}^q)$ alors la restriction de cette application à la sous variété $\mathrm{Sym}^k(\mathbb{C}^p \times \mathrm{Sym}^\ell(\mathbb{C}^q))$ fournit l'application m.

Définition 2

On dénotera par $\mathrm{Sym}^k(\mathbb{C}^p) \swarrow \mathrm{Sym}^\ell(\mathbb{C}^q)$ et on appellera $\mathrm{Sym}^k(\mathbb{C}^p)$ "accroche" à $\mathrm{Sym}^\ell(\mathbb{C}^q)$ la sous variété algèbrique de $\mathrm{Sym}^{\ell k}(\mathbb{C}^p \times \mathbb{C}^q)$ image de $\mathrm{Sym}^k(\mathbb{C}^p \times \mathrm{Sym}^\ell(\mathbb{C}^q))$ par l'application m de la proposition 1.

Proposition 2

La normalisation de $\mathrm{Sym}^k(\mathbb{C}^p) \swarrow \mathrm{Sym}^\ell(\mathbb{C}^q)$ est la variéte algèbrique $\mathrm{Sym}^k(\mathbb{C}^p \times \mathrm{Sym}^\ell(\mathbb{C}^q))$.

Démonstration

Soit $z \in \mathrm{Sym}^k(\mathbb{C}^p) \swarrow \mathrm{Sym}^\ell(\mathbb{C}^q)$ avec z écrit de la maniere suivante $z = ((x_1, y_1^1), \ldots \quad \ldots, (x_1, y_\ell^1), \ldots \quad \ldots, (x_k, y_1^k), \ldots$

...., $(x_k, y_\ell^k))$ où $x_i \neq x_j$ pour chaque $i \neq j$, alors il est facile de verifier que la fibre $m^{-1}(z)$ consiste d'un seul point.

Nous nous proposons de démontrer que au voisinage du point $z \in Sym^k(\mathbb{C}^p \times Sym^\ell(\mathbb{C}^q)) - m^{-1}(S)$ et $Sym^k(\mathbb{C}^p) \times Sym^\ell(\mathbb{C}^q) - S$ sont analytiquement isomorphes, pour cela on va à définir une application $\varphi$.

$$S = \{(x_1, y_1^1), \ldots, (x_1, y^1), \ldots, (x_k, y_\ell^k)/x_i = x_j, i \neq j\}$$

d'une voisinage de z dans l'accroche à valeurs dans $Sym^k(\mathbb{C}^p \times Sym^\ell(\mathbb{C}^q))$ cette application sera analytique et telle que $m \cdot \varphi = id$.

Comme tous les $x_i \in \mathbb{C}^p$ $i \in (1, 2, \ldots \ldots k)$ sont differents, il existent des voisinages $U_i$ de chaque $x_i$ dans $\mathbb{C}^p$ tels que leurs intersections sont vides, on dénote par $V_i^j$ un voisinage ouvert de $y_i^j$ dans $\mathbb{C}^q$.

Alors si $W = (U_1 \times V_1^1), \ldots \ldots, (U_k \times V_1^k) \ldots\ldots (U_k \times V_\ell^k)$, $p(W)$ est un voisinage ouvert de z dans l'accrroche. Pour définir $\varphi$ il suffira de la définir en Newton.

Soient $(M_1, \ldots \ldots, M_k)$ les fonctions de Newton des vecteurs $T_1, \ldots \ldots, T_k$ avec $T_i \in (\mathbb{C}^p \times Sym^\ell(\mathbb{C}^q))$ e'est à dire $T_i \in \mathbb{C}^p \times \bigoplus_1^\ell S_h(\mathbb{C}^q)$ où on a identifié $Sym^\ell(\mathbb{C}^q)$ avec son image par le plongement canonique.

Comme on connait les fonctions symétriques tensorielles de z on considère le polynôme :

$$P_z((\mu, \rho), t) = \sum_{k=0}^{\ell k} (-1)^{\ell k} s_h(\mu, \rho)\, t^{\ell k - h}$$

où $\mu$ et $\rho$ sont des formes linéaires sur $\mathbb{C}^p$ et $\mathbb{C}^q$ et $s_0 = 1$ respectivement.

$$T_j(z) = \frac{1}{2\pi i} \int_{|\mu(x_j)-t|=R_j} \frac{P'_z((\mu; \rho), t)}{P_z((\mu, \rho), t)} t^1 dt \;, \ldots\ldots\ldots,$$

$$\frac{1}{2\pi_i} \int_{|\mu(x_j)-t|=R_j} \frac{P'_z\,((\mu, \rho), t)}{P_z(((\mu, \rho), t)} t^\ell dt$$

On utilise maintenant le lemme suivant :

Lemme

Soient $(s_1, \ldots, \ldots, s_k) \in \overset{k}{\underset{1}{\oplus}} S_h(\mathbb{C}^p)$, u une forme linéaire sur $\mathbb{C}^p$ et $t \in \mathbb{C}$ $z = (x_1, \ldots, \ldots, x_k)$

$$P_z(u, t) = \sum (-1)^h S_h(u) \cdot t^{k-h}$$

Si le disque D(0, r) contient touts les racines de $P_z(u, t)$ alors on

$$N_m(z) = \frac{1}{2\pi_i} \int_{|t|=r} t^m \frac{P'_z(u, t)}{P_z(u, t)} dt$$

Démonstration

Voir B. Chap. 0 §2. lemme 2.

Comme le disque de centre $\mu(x_j)$ et de rayon $R_j$ contient seulement les racines :

$$(\mu(x_j), \rho(y_1^i)) \ldots \quad \ldots, (\mu(x_j), \rho(y^j)).$$

d'après le lemme on a que $T_j(z)$ nous donne les fonctions de Newton des vecteurs $(x_j, y_1^j), \ldots \ldots, (x_j, y_\ell^j)$. On définit alors l'application

$$\varphi(z) = (M_1(T_1, \ldots \quad \ldots, T_k), \ldots \ldots, M_k(T_1, \ldots \ldots, T_k))$$

où $M_i$ est la i-ème fonction symètrique de Newton, on obtient alors l'application cherchée.

Comme $Sym^k(\mathbb{C}^p)$ est normale, $Sym^k(\mathbb{C}^p \times Sym^\ell(\mathbb{C}^q))$ est aussi normale, l'unicité de la normalisation implique la proposition.

Remarques

Si $\ell = 1$ $Sym^k(\mathbb{C}^p \times Sym^\ell(\mathbb{C}^q)) \to \overset{m}{\to} \to\to Sym^{k\ell}(\mathbb{C}^p \times \mathbb{C}^q)$ est un isomorphisme et on a $Sym^k(\mathbb{C}^p \times Sym^\ell(\mathbb{C}^q)) \cong Sym^k(\mathbb{C}^p) \cong Sym^\ell(\mathbb{C}^q) \cong$ $\cong Sym^k(\mathbb{C}^p \times \mathbb{C}^q)$ neanmoins si $\ell > 1$ et même avec $p=q=k=1$, l'accroché est une sous variété stricte de $Sym^{k\ell}(\mathbb{C}^p \times \mathbb{C}^q)$, par exemple dans ce cas l'acrroché est une sous variété algèbrique de dimension complexe 3 de $Sym^2(\mathbb{C}^2)$ qui est une sous variété algèbrique de dimension complexe 4 de $\mathbb{C}^5$.

Avant d'énoncer la proposition qui va permettre de clasifier les revêtemenst ramifiés avec image directe donnée, nous preciserons quelques applications naturelles entre produits symétriques.

La restriction de l'application analytique :

$$p : Sym^{k\ell}(\mathbb{C}^p \times \mathbb{C}^q) \quad \to \quad Sym^{k\ell}(\mathbb{C}^p)$$

définie dans (G. §1. B.) induit par restriction une application analytique :

$$\tilde{p} : Sym^k(\mathbb{C}^p) \measuredangle Sym^\ell(\mathbb{C}^q) \to Sym^k(\mathbb{C}^p)$$

L'application de $(\mathbb{C}^p)^k$ dans $(\mathbb{C}^p)^{k\ell}$ définie par:

$$\ell(x_1, \ldots \ldots, x_k) = (x_1, \ldots \ldots, x_k, \ldots \ldots, x_1, \ldots, x_k)$$

passe au quotient pour définir une application analytique :

$$\ell : Sym^k(\mathbb{C}^p) \to Sym^{k\ell}(\mathbb{C}^p)$$

Nous rappellons maintenant une proposition qui sera fondamentale dans la démonstration de la proposition 4.

Proposition 3

Soit U un polydisque de $\mathbb{C}^n$, B un polydisque de $\mathbb{C}^p$ où bien $\mathbb{C}^p$. Il y a une bijection naturelle entre l'ensemble des revêtements ramifiés de degré k de U contenus dans U × B et l'ensemble des applications analytiques $f : U \to Sym^k(B)$.

Démonstration.

Voir B. Chapitre 0 Proposition 3.

Proposition 4

Il y a une bijection naturelle entre les revêtements ramifiés de degré $k\ell$ de U contenus dans $U \times \mathbb{C}^p \times \mathbb{C}^q$ avec image directe $\ell \cdot X$ dans $U \times \mathbb{C}^p$ (où X est un revêtement ramifié de degré k sans multiplicités de U contenu dans $U \times \mathbb{C}^p$) et les applications analytiques F qui font conmutative le diagramme suivant :

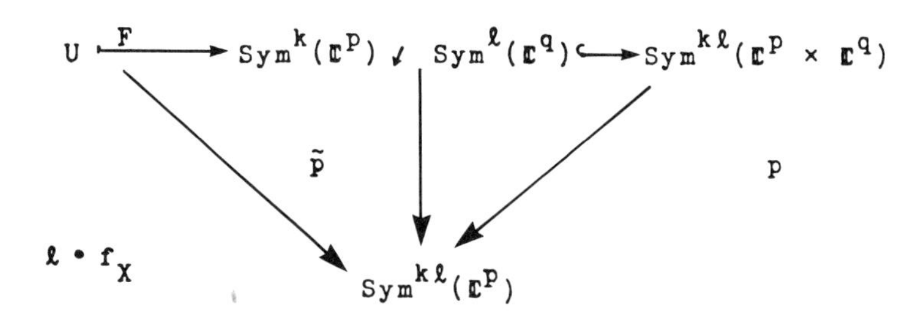


Démonstration

Soit $F : U \to Sym^k(\mathbb{C}^p) \swarrow Sym^\ell(\mathbb{C}^q)$ une application analytique qui fait conmutative le diagramme, alors F est une application dans $Sym^{k\ell}(\mathbb{C}^p \times \mathbb{C}^q)$ à valeurs dans la sous variété fermée l'accroché par la proposition 3. On peut lui associer un revêtement ramifié de degré $k\ell$ de U contenu dans $U \times \mathbb{C}^p \times \mathbb{C}^q$ qu'on notera par $Y_F$, par définition d'image directe de $Y_F$ par projection partielle et à cause de la conmutativité du diagramme l'image directe de $Y_F$ est alors $\ell \cdot X$.

Soit $Y \subset U \times \mathbb{C}^p \times \mathbb{C}^q$ un revêtement ramifié de degré $k\ell$ de U avec image direct $\ell \cdot X$ dans $U \times \mathbb{C}^p$, on dénotera par $R(|X|)$ et $R(|Y|)$ les ramifications de $|X|$ et $|Y|$ respectivement, les deux sont des sous ensembles analytiques fermés d'interieur vide dans U.

Pour chaque $t \in U - R(|X|) \cup R(|Y|)$ il existe un voisinage $V_t$ où sont définies les branches locales de Y

$$g_j^i : V_t \to \mathbb{C}^p \times \mathbb{C}^q \quad i \times j \in (1,\ldots,k)\times(1,\ldots,\ell)$$

on peut supposer $(g_j^i) = (x_j^i, y_j^i) \in \mathbb{C}^p \times \mathbb{C}^q$, comme on sait que l'image directe de Y est $\ell \cdot X$ il y aura pour chaque $t' \in V_t$ k branches $x_j^i(t')$ differents, comme les branches locales $g_j^i$ sont définies modulo une permutation de $S_{k\ell}$ on peut mettre sous la forme :

$$((x_1^1(t'), y_1^1(t'), \ldots \ldots, (x_\ell^k(t'), y_\ell^k(t'))$$

$$\text{avec } x_1^1(t') = \ldots = x_\ell^1(t'), \ldots \ldots, x_1^k(t') = \ldots = x_\ell^k(t')$$

$$\text{on définit alors } u \circ (g_j^i) : U-R(|X|) \cup R(|Y|) \longmapsto \bigoplus_1^k S_h(\mathbb{C}^p \times \mathbb{C}^q)$$

où u est le plongement canonique de $\mathrm{Sym}^{k\ell}(\mathbb{C}^p \times \mathbb{C}^q)$ dans l'espace affin $\bigoplus_1^{k\ell} S_h(\mathbb{C}^p \times \mathbb{C}^q)$, comme $\Pi : |Y| \to U$ est propre cette avant dernière application est localement bornée sur

$R(|X|) \cup R(|Y|)$, alors elle se prolonge à une application analytique :

$$F : U \to \overset{k}{\underset{1}{\oplus}} S_h(\mathbb{C}^p \times \mathbb{C}^q)$$

et les valeurs de ce prolongement restent dans $u(\mathrm{Sym}^k(\mathbb{C}^p) \swarrow \mathrm{Sym}^\ell(\mathbb{C}^q))$ par continuité. On définit alors le prolongement de $u \circ (g_j^i)$ par composition avec $u^{-1}$, il est claire de la construction de F, la conmutativité du diagramme.

## §2. DEFORMATIONS DE REVÊTEMENTS RAMIFIÉS

Soit $D = \{s \in \mathbb{C} \ / \ |s| < 1\}$, U un polydisque de $\mathbb{C}^n$ une famille analytique des revêtements ramifiés de degré k de U contenus dans $U \times \mathbb{C}^p$ parametrés par le disque unite D est une application analytique : $\phi : D \times U \to \mathrm{Sym}^k(\mathbb{C}^p)$.

Définition 1

Soit X un revêtement ramifié de degré k de U contenu dans $U \times \mathbb{C}^p$, nous appellerons deformation de X avec base D la donné d'une famille analytique des revêtements ramifiés de degré k de U contenus dans $U \times \mathbb{C}^p$ parametrée par D, telle que $\phi(o, t) = X$.

Exemples

Si $X = (x_1, \ldots \ \ldots, x_k) \in \mathrm{Sym}^k(\mathbb{C}^p)$ alors une deformation de X n'est qu'une application analytique:

$$\phi : D \to \mathrm{Sym}^k(\mathbb{C}^p)$$

telle que $\phi(0) = (x_1, \ldots \ \ldots, x_k)$.

Soit $U = \mathbb{C}$ et l'application analytique:

$$\phi : D \times \mathbb{C} -\{0\} \to \mathbb{C} \times \mathbb{C} \to \bigoplus_1^2 S_h(\mathbb{C})$$

$$(s, t) \to (s + \sqrt{t}, s - \sqrt{t}) \to (2s, s^2-t)$$

elle se prolonge dans une deformation du revêtement ramifié associe à $\sqrt{t}$.

Maintenant on va montrer que la donnée d'une application de U à valeurs dans l'accroché de $Sym^k(\mathbb{C}^p)$ avec lui même permet construire une deformation canonique appellé affine d'un revêtement ramifié avec multiplicités de U contenu dans $U \times \mathbb{C}^p$.

Pour cela comme d'habitude on obligé à expliciter quelques applications naturelles entre les produits symétriques.

Soit $s \in \mathbb{C}$, l'application de $(\mathbb{C}^p \times \mathbb{C}^q)^k$ dans lui-même définie par $(x_1, y_1) \ldots (x_k, y_k) \to (x_1, sy_1) \ldots (x_k, sy_k)$ induit une application analytique:

$$Sym^k(id \times s) : Sym^k(\mathbb{C}^p \times \mathbb{C}^q) \to Sym^k(\mathbb{C}^p \times \mathbb{C}^q)$$

Aussi l'application de $(\mathbb{C}^p \times \mathbb{C}^p)^k$ dans $(\mathbb{C}^p)^k$ induit une autre application analytique:

$$Sym^k(+) : Sym^k(\mathbb{C}^p \times \mathbb{C}^p) \to Sym^k(\mathbb{C}^p)$$

<u>Proposition 1</u>

Soit $X_o$ un revêtement ramifié de degré k de U contenu dans $U \times \mathbb{C}^p$ tel que $fX_o(0) = k\cdot\{0\}$. Pour chaque $F : U \to Sym^k(\mathbb{C}^p) \measuredangle Sym^\ell(\mathbb{C}^q)$ tel que $p_*F = \ell\cdot X_o$ il existe une deformation canonique de $\ell\cdot x_o$.

$$\tilde{F} : D \times U \to Sym^{k\ell}(\mathbb{C}^p).$$

Démonstration

Soit $F : U \to \mathrm{Sym}^k(\mathbb{C}^p) \not\subset \mathrm{Sym}^\ell(\mathbb{C}^q)$ telle que on ait $F(t) = (x_i(t), y_i^j(t))$ $i \in I$, $j \in J$ (I a k éléments et J a éléments). On définit $F' : D \times U \to \mathrm{Sym}^{k\ell}(\mathbb{C}^p \times \mathbb{C}^q)$ de la manière suivante: $F'(s,t) = (x_i(t), sy_i^j(t))$. On considère les applications analytiques ci dessus et on met $\tilde{F} = \mathrm{Sym}^{kl}(+) \circ \mathrm{Sym}^k (id \times s) \circ F'$.

On obtient alors $\tilde{F} : U \times D \to \mathrm{Sym}^{\ell k}(\mathbb{C}^p)$ analytique et $\tilde{F}(0, t) = \ell \cdot X_o$.

Définition 2

On appellera deformation "affine" associé à $\ell \cdot X_o$ la deformation construite dans la proposition 1.

Remarque

Si $\phi : U \times D \to \mathrm{Sym}^{k\ell}(\mathbb{C}^p)$ est une deformation de $\ell \cdot X_o$ telle que $\phi(s, o) = \ell k \cdot \{0\}$ en imposant des conditions assez fortes à la deformation $\phi$ on peut construire un élement:

$$F_\phi : U \to \mathrm{Sym}^k(\mathbb{C}^p) \not\subset \mathrm{Sym}^\ell(\mathbb{C}^q)$$

tel que la deformation affine $\tilde{F}_\phi$ soit tangente à $\phi$ en $0 \in D$ le long de U.

Ces conditions : majoration de la derivée $\frac{\partial \phi}{\partial s}(0,t)$ et non annullation de cette derivée correspondent respectivement à la notion de revêtement ramifié tranverse à l'origine de Henault et la non annullation correspond à la non existence de "couches multiples" de Barlet.

L'auteur espère pouvoir donner des conditions géometriques explicites dans la suite de ce travail.

BIBLIOGRAPHIE

B. Barlet D. L'espace des cycles ....
Séminaire F. Norguet Lectures Notes 482.

B. Barlet D. Deformation à l'ordre 1 des Cycles
Prepublication de L'institut Elie Cartan
Nancy.

G. González-Aguilera V. Cycles analytiques compacts d'un
fibré vectoriel à apparaitre dans le Séminaire
F; Norguet 5 Lec; Notes in Math.

H. Henault A. Cycle tangente de Zariski dans le Séminaire
F. Norguet 4. Lec. Notes in Math.

# 16
# Asymptotical Statistical Inference and Gaussian Processes

GUIDO E. DEL PINO / Departamento de Probabilidad y Estadística, Pontificia Universidad Católica de Chile, Santiago, Chile

## 1. INTRODUCTION AND THE MAIN IDEA

In this paper we study the asymptotic local behavior of a family $(Q_{n\theta}, \theta \in \Theta)$ of probability distributions near the point $\theta_0 \in \Theta$ and when n tends to infinity. One typical case is when n represents the sample size, $\theta$ is an unknown parameter and one wants to study tests for the null hypothesis $\theta = \theta_0$. The families $(Q_{n\theta}, \theta \in \Theta)$ need not belong to the same sample space for different values of n. The method used to study this problem is as follows:

Through a transformation of the sample space we consider stochastic processes $\eta_n(\cdot,\theta_0)$ with induced probability measures $P_n(\cdot,\theta_n,\theta_0)$ in some space of functions. We assume that $\Theta$ is a subset of a topological vector space H (typically the Euclidean space $\mathbb{R}^k$) and the stochastic processes are chosen in such a way that there exists a sequence of constants d(n) so that

$$\lim_n d(n)(\theta_n - \theta_0) = \beta \tag{1.1}$$

implies that $P_n(\cdot,\theta_n,\theta_0)$ converges in some sense (finite dimensional distributions, weakly, etc.) to a probability measure $(P_\beta(\cdot,\theta_0))$. The main idea of this paper is to study the asymptotic local behavior of $(Q_{n\theta}, \theta \in \Theta)$ near $\theta_0$, on the basis of the family of probability measures $(P_\beta(\cdot,\theta_0), \beta \in H)$.

To study the nearness of the probability distributions $Q_{n\theta_n}$ and $Q_{n\theta_0}$ we consider a sequence of level $\alpha$ test $\theta = \theta_0$ vs. $\theta = \theta_n$ based on $\eta_n(\cdot,\theta_0)$. Let $\gamma_n(\alpha,\theta_0,\theta_n)$ be the corresponding powers of the tests and let

$$\lim_n \gamma_n(\alpha,\theta_0,\theta_n) = \tilde{\gamma}(\alpha,\theta_0,\beta) \tag{1.2}$$

Now, let $\eta(\cdot,\theta_0)$ be an stochastic process with distribution $P_\beta(\cdot,\theta_0)$ and consider the test

$$\beta = 0 \quad \text{vs.} \quad \beta = c$$

based on $\eta(\cdot,\theta_0)$.

Let $\gamma(\alpha,\theta_0,c)$ be the power of this test. There are important examples where

$$\gamma(\alpha,\theta_0,\beta) = \tilde{\gamma}(\alpha,\theta_0,\beta)$$

Then the computation of the limit (1.2), which is usually difficult may be replaced by the easier problem of finding $\gamma(\alpha,\theta_0,\beta)$. The computations are particularly easy when $P_\beta(\cdot,\theta_0)$ is a Gaussian measure with covariance function independent of $\beta$ and mean function linear in $\beta$. The general theory for this situation may be found, for instance, in del Pino (1979).

In Section 2 we consider the case of empirical processes and in Section 3 we consider the case of quantile processes. Finally, in Section 4 we present a situation where the general method of this paper fails.

## 2. THE CASE OF EMPIRICAL PROCESSES

Let $X_1, \ldots, X_n$ be independent and identically distributed with distribution function $F(\cdot,\theta)$, where $\Theta$ is an open set in $\mathbb{R}^k$. Here $Q_{n\theta}$ is the product measure $(F(\cdot,\theta))^n$ defined on $\mathbb{R}^n$.

The empirical distribution function (e.d.f.) defined by

$$F_n(x) = \frac{1}{n} \quad (\text{number of } X_i \le x) \qquad -\infty < x < \infty$$

is a sufficient statistic, so that it contains all relevant information for inferences about $\theta$. For a fixed $\theta_0 \in \Theta$ define the empirical processes $\eta_n(\cdot,\theta_0)$ by

$$\eta_n(x,\theta_0) = n^{1/2}(F_n(x) - F(x,\theta_0)) \qquad -\infty < x < \infty$$

The empirical process $\eta_n(\cdot,\theta_0)$ induces a probability measure $P_n(\cdot,\theta)$ on the Skorokhod space $D(-\infty,\infty)$ of right continuous functions over $\mathbb{R}$, with left hand limits. The finite dimensional distributions (f.d.d.) of $P_n(\cdot,\theta_0)$ converge to those of or Gaussian measure $P_0(\cdot,\theta_0)$ with mean zero and covariance function $K_{\theta_0}(x,y)$, where

$$K_\theta(x,y) = \min(F(x,\theta),F(y,\theta)) - F(x,\theta)F(y,\theta) \tag{2.1}$$

Consider now the following sequence of parameters:

$$\theta_n = \theta_0 + \beta n^{-1/2}$$

It may be easily proved [see del Pino (1980)] that if $F(\cdot,\theta)$ is differentiable with partial derivatives $F_i(\cdot,\theta)$, $i = 1, \ldots, k$, then

$$P_n(\cdot,\theta_n) \xrightarrow{\text{f.d.d.}} P_\beta(\cdot,\theta_0)$$

where $P_\beta(\cdot,\theta_0)$ is a Gaussian measure with covariance function (2.1) and mean function

$$m_\beta(x,\theta_0) = \sum_{i=1}^{k} \beta_i F_i(x,\theta_0) \tag{2.2}$$

Let $F(\cdot,\theta)$ have a density $f(\cdot,\theta)$ and assume that $f(x,\cdot)$ has continuous partial derivatives. Let

$$\ell_i(x,\theta) = \frac{\partial \log f(x,\theta)}{\partial \theta_i} \qquad i = 1, \ldots, k \tag{2.3}$$

The locally most powerful test of

$$\theta = \theta_0 \quad \text{vs.} \quad \theta = \theta_n = \theta_0 + cn^{-1/2}$$

rejects the null hypothesis for large values of

$$T_n = n^{-1/2} \sum_{i=1}^{k} c_i \sum_{i=1}^{n} \ell_i(x_i,\theta_0) \tag{2.4}$$

It is easy to find (see, e.g., Cox and Hinkley (1974), pp. 106-109) that

$$E_{\theta_0} T_n = 0 \qquad \mathrm{Var}_{\theta_0} T_n = c'I(\theta_0)c$$

$$E_{\theta_n} T_n = c'I(\theta_0)c + o(1) \qquad \mathrm{Var}_{\theta_n} T_n = c'I(\theta_0)c + o(1)$$

The matrix $I(\theta_0)$ is a Fisher information matrix for one observation and

$$I_{ij}(\theta_0) = E_{\theta_0} \ell_i(X_1)\ell_j(X_1) \qquad i,j = 1, \ldots, k$$

Furthermore, for large n, $T_n$ is asymptotically normal and the asymptotic power $\tilde{\gamma}(\alpha,\theta_0,c)$ of these tests is that of

$$Y \sim N(0,\sigma^2) \quad \text{vs.} \quad Y \sim N(\sigma^2,\sigma^2)$$

where $\sigma^2 = c'I(\theta_0)c$. Then

$$\tilde{\gamma}(\alpha,\theta_0,c) = \phi(\phi^{-1}(\alpha) + \sigma) \tag{2.5}$$

On the other hand the power of the test $P = P_0$ vs. $P = P_\beta$, based on a realization of the Gaussian process $\eta$ is

$$\gamma(\alpha,\theta_0,\beta) = \phi(\phi^{-1}(\alpha) + \|m_\beta(\cdot,\theta_0)\|) \tag{2.6}$$

where the norm in (2.6) is computed in the sense of the reproducing kernel Hilbert space with kernel K.

It may be proved [see, e.g., del Pino (1979)] that

$$\langle u,v\rangle = \int \left(\frac{u'(x)}{f(x,\theta)}\right)\left(\frac{v'(x)}{f(x,\theta)}\right) f(x,\theta)\, dx$$

Hence

$$\langle F_i(\cdot,\theta_0)F_j(\cdot,\theta_0)\rangle = I_{ij}(\theta_0)$$

where we have used the fact that

$$\frac{d}{dx}\left(\frac{\partial F}{\partial\theta_i}(x,\theta_0)\right) = \frac{\partial}{\partial\theta_i} f(x,\theta_0)$$

Then

$$\begin{aligned}\|m_\beta(\cdot,\theta_0)\|^2 &= \sum_{i=1}^{k}\sum_{j=1}^{k} \beta_i\beta_j I_{i'}(\theta_0) \\ &= \beta' I(\theta_0)\beta\end{aligned}$$

Hence

$$\gamma(\alpha,\theta_0,\beta) = \tilde{\gamma}(\alpha,\theta_0,\beta)$$

Furthermore, the optimal test of $P_0$ vs. $P_\beta$ rejects the null hypothesis for large values of

$$\sum_{i=1}^{k} c_i \int \ell_i(\cdot,\theta_0)d\eta(\cdot,\theta_0)$$

This suggests that for testing $\theta = \theta_0$ vs. $\theta = \theta_n$ one should reject the null hypothesis for large values of

$$V_n = \sum_{i=1}^{k} c_i \int \ell_i(\cdot,\theta_0)d\eta_n(\cdot,\theta_0)$$

$$= n^{-1/2} \sum_{i=1}^{k} c_i \sum_{j=1}^{k} \ell_i(x_j,\theta_0)$$

which coincides with $T_n$ defined in (2.4).

## 3. THE CASE OF QUANTILE PROCESSES

Let $X_1, \ldots, X_n$ be i.i.d. $F(\cdot,\theta)$, where $F(\cdot,\theta)$ is a continuous distribution function. Let $X_{1:n} < X_{2:n} < \cdots < X_{n:n}$ be the corresponding order statistic. Assume that $F(\cdot,\theta)$ has a continuous positive derivative $f(\cdot,\theta)$ and define the *quantile function* $Q(\cdot,\theta)$ by

$$F(Q(t,\theta)) = t \qquad 0 < t < 1 \tag{3.1}$$

Then $Q(\cdot,\theta)$ also has a continuous positive derivative $q(\cdot,\theta)$ and

$$q(\cdot,\theta) = (f(Q(\cdot,\theta),\theta))^{-1} \tag{3.2}$$

Define the empirical quantile function $Q_n(\cdot)$ by

$$Q_n(t) = X_{i:n} \qquad \text{for } (i-1)/n < t \le i/n \qquad i = 1, \ldots, n \tag{3.3}$$

Finally, define the quantile process $\psi_n(\cdot,\theta)$ by

$$\psi_n(t,\theta) = n^{1/2}(Q_n(t) - Q(t,\theta)) \qquad 0 < t < 1 \tag{3.4}$$

and let $P_n(\cdot,\theta,\theta_0)$ be the probability measure induced by $\psi_n(\cdot,\theta_0)$. If $\theta_n = \theta_0 + cn^{-1/2}$ then, under further smoothness assumptions on $Q(\cdot,\cdot)$ one gets

$$P_n(\cdot,\theta_n,\theta_n) \xrightarrow{\text{f.d.d.}} P_0(\cdot,\theta_0) \tag{3.5}$$

where $P_0(\cdot,\theta)$ is a Gaussian probability measure with mean zero and covariance function

$$G_\theta(s,t) = q(s,\theta)q(t,\theta)(\min(s,t) - st) \qquad 0 < s,t < 1 \tag{3.6}$$

Assuming $Q(t,\cdot)$ to be differentiable at $\theta_0$, then from

$$\psi_n(\cdot,\theta_0) = \psi_n(\cdot,\theta_n) + n^{1/2}(Q(t,\theta_n) - Q(t,\theta_0))$$

it follows that

$$P_n(\cdot,\theta_n,\theta_0) \xrightarrow{\text{f.d.d.}} P_c(\cdot,\theta_0), \tag{3.7}$$

where $P_c(\cdot,\theta_0)$ is a Gaussian probability measure with covariance function $G_{\theta_0}$ and mean function

$$\sum_{i=1}^{k} c_i Q_i(\cdot,\theta_0), \tag{3.8}$$

where $Q_i(t,\cdot)$ is the ith partial derivative of $Q(t,\cdot)$.

The covariance function (3.6) is not very convenient to work with. For this reason we define a modified quantile process $\eta_n(\cdot,\theta)$ by

$$\eta_n(\cdot,\theta) = \psi_n(\cdot,\theta)/q(\cdot,\theta) \tag{3.9}$$

If $q(t,\cdot)$ is continuous then, under $\theta = \theta_n$, the f.d.d. of $\eta_n(\cdot,\theta_0)$ converge to those of a Gaussian process with covariance function K and mean function $m_c(\cdot,\theta_0)$, where

$$m_c(\cdot,\theta) = \sum_{i=1}^{k} c_i Q_i(\cdot,\theta)/q(\cdot,\theta) \tag{3.10}$$

and

$$K(s,t) = \min(s,t) - st \qquad 0 < s,t < 1 \tag{3.11}$$

A simple computation shows that $m_c(\cdot,\theta)$ has the equivalent expression

$$m_c(\cdot,\theta) = -\sum_{i=1}^{k} c_i F_i(Q(\cdot,\theta),\theta) \tag{3.12}$$

The Hilbert space H(K) is the set of all functions h such that

$$h(t) = \int_0^t g(u)\,du \qquad 0 < t < 1 \tag{3.13}$$

for some function g satisfying

$$\int_0^1 g(u)\,du = 0 \tag{3.14}$$

$$\int_0^1 g^2(u) < \infty \tag{3.15}$$

The function g may be denoted by h'. Then the inner product of $h_1$ and $h_2$ in H(K) is

$$<h_1,h_2>_{H(K)} = \int_0^1 h_1'(t)h_2'(t)\,dt$$

Assume $f(x,\cdot)$ to have an ith partial derivative $f_i(x,\cdot)$ for all x, and that $F_1'(x,\theta) = f_i(x,\theta)$ for all $x,\theta$. Then, the change of variable $t = F(x,\theta)$ yields

$$\begin{aligned} A_{ij}(\theta) &= <F_iQ(\cdot,\theta),\theta),F_j(Q(\cdot,\theta),\theta)>_{H(K)} \\ &= \int f_i(\cdot,\theta)f_j(\cdot,\theta)/f(\cdot,\theta) \qquad i = 1, \ldots, n \end{aligned}$$

so that the $A(\theta)$ coincides with Fisher's information matrix for $X_1 \sim F(\cdot,\theta)$.

Any random variable in $L_2(\eta(\cdot,\theta))$ may be represented in the form

$$L_g(\eta(\cdot,\theta)) = \int_0^1 g(t,\theta)d\eta(t,\theta) \tag{3.16}$$

where $g(\cdot,\theta)$ satisfies (3.14) and (3.15). In this form, however, it is difficult to give a useful meaning to $L_g(\eta_n(\cdot,\theta))$. Assuming that $g(\cdot,\theta)$ is differentiable and that $g(\cdot,\theta)$ approaches 0 quickly as $t \to 0$ or $t \to 1$, it is possible to integrate (3.16) by parts to get

$$L_g(\eta(\cdot,\theta)) = -\int_0^1 g'(t,\theta)\eta(t,\theta)\,dt \tag{3.17}$$

Let

$$a(t,\theta) = -g'(t,\theta)/q(\theta) \tag{3.18}$$

Then

$$L_g(\eta_n(\cdot,\theta)) = n^{1/2}\left\{\int_0^1 a(t,\theta)Q_n(t)\,dt - \int_0^1 a(t,\theta)Q(t,\theta)\,dt\right\} \tag{3.19}$$

which is well approximated by

$$W_n(a,\theta) = n^{1/2}\left\{\frac{1}{n}\sum_{i=1}^{n} a(i/n,\theta)X_{i:n} - \int_0^1 a(t,\theta)Q(t,\theta)\,dt\right\} \qquad (3.20)$$

Thus the linear functionals of the processes $\eta_n(\cdot,\theta)$ are closely related to linear functions of order statistics.

The function $a(\cdot,\theta)$ corresponding to (3.12) may be proven to be

$$a(t,\theta) = \sum_{i=1}^{k} c_i \ell_i'(Q(t,\theta),\theta) \qquad (3.21)$$

where $\ell_1'(\cdot,\theta)$ is the derivative of the function $\ell_i(\cdot,\theta)$ given by (2.3).

## 4. A COUNTEREXAMPLE TO THE GENERAL METHOD

Let $X_1$, $X_2$, ... be a sequence of i.i.d. random variables with mean $\mu(\theta)$ and variance $\sigma^2(\theta)$, with $\theta \in \Theta$, an open set in $\mathbb{R}^k$. Consider the sequence of stochastic processes $(Y_n(\cdot) = (Y_n(1), Y_n(2), \ldots)$, $n = 1, 2, \ldots$, defined by

$$\begin{aligned} Y_n(s) &= X_s \qquad s = 1, \ldots, n \\ &= 0 \qquad s < n \end{aligned} \qquad (4.1)$$

For a fixed value $\theta = \theta_0$ define $\eta_n(\cdot,\theta_0) = (\eta_n(t,\theta_0),\ 0 \le t \le 1)$ by

$$\eta_n(t,\theta_0) = n^{-1/2}\left\{\sum_{i=1}^{[nt]} X_i - n\mu(\theta_0)t\right\} \qquad 0 \le t \le 1 \qquad (4.2)$$

Assume $\theta_n = \theta_0 + \beta n^{-1/2}$. If $\mu(\cdot)$ is a continuously differentiable function and $a(\theta_0)$ is the column vector of partial derivatives of $\mu(\cdot)$ at $\theta = \theta_0$,

$$\lim_n n^{1/2}(\mu(\theta_n) - \mu(\theta_0)) = a(\theta_0)^t\beta \qquad (4.3)$$

From (4.2) we may write

$$\eta_n(t,\theta_0) = \eta_n(t,\theta_n) + tn^{1/2}(\mu(\theta_n) - \mu(\theta_0))$$

Under $P_n(\cdot,\theta_n)$ the f.d.d. of $\eta_n(\cdot,\theta_n)$ have the same limit as those of $\eta_n(\cdot,\theta_0)$ under $P(\cdot,\theta_0)$. From this it follows that $P_n(\cdot,\theta_n)$ converges to a Gaussian measure $P_\beta(\cdot,\theta_0)$ with mean function

$$m_\beta(t,\theta_0) = (a(\theta_0)^t\beta)t \qquad 0 \le t \le 1 \qquad (4.4)$$

and covariance function

$$K_{\theta_0}(t_1,t_2) = \sigma^2(\theta_0)\ \min(s,t) \qquad 0 \le t_1,t_2 \le 1 \tag{4.5}$$

We may note that

$$m_\beta(\cdot,\theta_0) = \frac{a^t\beta}{\sigma^2(\theta_0)} K_{\theta_0}(\cdot,\ 1)$$

and from this it follows that the most powerful test of $\beta = 0$ vs. $\beta = c$ based on $n(\cdot,\theta_0)$ rejects $\beta = 0$ for large (small) values of $n(1,\theta_0)$ if $a^t(\theta_0)c > 0$ ($< 0$). We have

$$|m_c(\cdot,\theta_0)| = |a(\theta_0)^t c|/\sigma(\theta_0) \tag{4.6}$$

It follows from (4.6) that if $a^t c = 0$ the power of the test just mentioned is equal to $\alpha$. On the other hand, the implied test for finite sample size would reject $\theta = \theta_0$ for large (small) values of $\eta_n(1,\theta_0)$ if $a^t c > 0$ ($< 0$). But $\eta_n(1,\theta_0) = n^{1/2}(\bar{X} - \mu(\theta_0))$ so that the test would reject $\theta = \theta_0$ in favor of $\theta = \theta_0 + cn^{-1/2}$ for large (small) values of the mean $\bar{X}$ if the angle between c and $a(\theta_0)$ is less (more) than 90 degrees. It is clear that the asymptotic relative efficiency with respect to the best test may be arbitrarily close to zero and it is equal to zero if c and $a(\theta_0)$ are orthogonal.

## ACKNOWLEDGMENT

This work has been partially supported by Dirección de Investigación, Pontificia Universidad Católica de Chile, Project 42-82.

## REFERENCES

D. R. Cox and D. V. Hinkley, Theoretical Statistics, Chapman and Hall, London, 1974.

G. E. del Pino, Statistical inference for Gaussian processes with known covariance function, Univ. of Wisconsin, Dept. of Statistics, T. R. No. 599, 1979.

G. E. del Pino, Asymptotic statistical inference based on asymptotic Gaussian processes, Univ. of Wisconsin, Dept. of Statistics, T. R. No. 628, 1980.

# 17
# Sur les Methodes Asymptotiques de l'Analyse Stochastique

ROLANDO REBOLLEDO / Departamento de Matemática, Pontificia Universidad Católica de Chile, Santiago, Chile

Ces dernières années ont connu l'essor du Calcul Stochastique, ceuvre en chantier où confluent les recherches en Théorie Générale de Processus, Théorie de Martingales, Intégrales Stochastiques.

D'autre part, dans les années 50, la Théorie de Processus était marquée par les recherches sur les Topologies Faibles pour les familles de mesures définies sur les espaces métriques de fonctions.

La Théorie de Martingales a mis en relief des résultats de "structure" très beaux et profonds: l'étude de l'intégrale stochastique a été possible grâce aux théorèmes de décomposition de surmartingales, de martingales locales; grâce a la découverte des espaces estables et à l'introduction des espaces $H^p$. On a ainsi progressé jusqu'au résultat fondamental de Dellacherie que caractérise la classe la plus vaste des processus par rapport auxquels on peut "raisonnablement" définir une "intégrable": c'est la classe de semimartingales.

---

Ce travail a été réalisé avec le concours partiel de la Dirección de Investigación de la Universidad Católica de de Chile (DIUC).

Du côté Topologies Faibles, l'accent avait été mis sur la caractérisation des processus du point de vue de leur loi, conçues comme des mesures particulières définies sur des espaces de fonctions bien précis.

Il était naturel de penser à lier ces deux lignes de recherche: c'est-à-dire, d'étudier d'une part l'intervention des topologies faibles dans le Calcul Stochastique et, d'autre part, d'appliquer les résultats du calcul stochastique à certains problèmes de Topologie liés aux lois des processus.

Le but du présent article est de résumer brièvement les principaux résultats de l'auteur dans cette dernière direction de recherche.

## §1. INTRODUCTION / GENERALITES

Dans les cours élémentaires de Probabilités on apprend le résultat suivant: étant donné une suite $(\xi_n)_{n\geq 1}$ de variables aléatoires indépendantes, de même loi, de moyenne nulle et de variance $\sigma^2>0$ finie, la suite $(X_n)_{n\geq 1}$ de variables aléatoires définie par

$$X_n := \frac{\xi_1 + \ldots + \xi_n}{\sigma\sqrt{n}} , \qquad (n\in\mathbb{N}^*)$$

converge en loi vers une variable aléatoire normale centrée réduite. C'est le Théorème de la Limite Centrale.

Donsker généralisa ce résultat vers les années 50, en l'étendant aux processus de la manière suivante. Si l'on définit pour tout $t\in\mathbb{R}_+$ et tout $n\in\mathbb{N}$, $M_n(t) := X_{[nt]}$, où $[\cdot]$ désigne la partie entière de $\cdot$, alors la suite de processus $(M_n)_{n\geq 1}$ converge en loi vers un mouvement brownien canonique. C'est le Principe d'Invariance.

Le terme "convergence en loi" du Principe d'Invariance signifie la convergence étroite (ou faible) des lois $(P_n)_{n\geq 1}$

associées aux processus $(M_n)_{n\geq 1}$ sur un espace métrique particulier: l'espace $D = D(\mathbb{R}_+ , \mathbb{R})$ des fonctions continues à droite et limitées à gauche de $\mathbb{R}_+$ dans $\mathbb{R}$ , muni de la topologie de Skorokhod (BILLINGSLEY [1]).

Par ailleurs, les variables $(\xi_n)$ sont supposées être définies sur un espace probabilisé de base $(\Omega, F, \mathbb{P})$. Sur cet espace nous introduisons les tribus $F_{nt} := \sigma(\xi_k;\ k\leq [nt])$ $(n\in\mathbb{N}\ ,\ t\in\mathbb{R}_+)$. Appelons $\mathbb{F}_n := (F_{nt}\ ;\ t\in\mathbb{R}_+)$ la filtration qui en résulte pour tout $n\in\mathbb{N}$. Avec ce choix de filtration, chaque $M_n$ devient une martingale par rapport à $\mathbb{F}_n$ et $\mathbb{P}$. Mieux encore, $M_n$ est une martingale de carré intégrable et si l'on cherche son processus croissant associé $< M_n, M_n>$ (celui qui est croissant, prévisible, et qui fait que $M_n^2 - <M_n, M_n>$ soit également une martingale) on trouve qu'il vaut

$$<M_n, M_n> = \frac{[nt]}{n}\ ,\quad (n\in\mathbb{N}\ ,\ t\in\mathbb{R}_+).$$

D'autre part, le mouvement brownien canonique, soit $M$, est une martingale continue dont le processus croissant associé $<M, M>$ vaut $<M, M>(t) = t$ $(t\in\mathbb{R}_+)$. On remarquera que la suite $(<M_n, M_n>(t))$ converge vers $<M, M>(t)$ pour tout $t\in\mathbb{R}_+$. La question que l'on se pose est de savoir dans quelle mesure cette dernière convergence entraîne la convergence en loi de $(M_n)$ vers $M$. C'est un exemple du type de problèmes que nous voulons aborder.

En outre, le Théorème de Prokhorov occuppe une place fondamentale dans l'étude de la Topologie Etroite: il permet de caractériser les familles de mesures de probabilité relativement étroitement compactes. Nous étudiérons ci-dessous les liens entre ces notions et les temps d'arrêt. C'est l'autre type de problèmes que nous aborderons.

Dans ce qui suit nous nous servirons sans autre explication de notions du Calcul Stochastique telles qu'elles figurent par exemple dans DELLACHERIE-MEYER [1],[2] et JACOD [1]. Pour les topologies faibles les références de base sont BILLINGSLEY [1] et PARTHASARATHY [1].

1.

Tout le long de cet article nous considérons un espace probabilisé complet $(\Omega, F, \mathbb{P})$. Sur cet espace nous allons prendre différents processus et différentes filtrations. Nous nous intéressons aux processus dont les trajectoires sont continues à droite et limitées à gauche (càdlàg), c'est pourquoi le terme "processus" désigne ci-dessous une application mesurable $X : \Omega \to D$ où $D$ est l'espace de Skorokhod $D(\mathbb{R}_+ R)$ muni de sa tribu borélienne $\mathcal{D}$. Si nous ajoutons une filtration $\mathbb{F} = (F_t;\ t\in\mathbb{R}_+)$ sur $(\Omega, F, \mathbb{P})$ et que X lui est adapté, nous parlerons alors du "processus" $(X, \mathbb{F})$ pour mettre en relief l'adaptation.

Passons maintenant à l'espace $(D, \mathcal{D})$: nous y considérons les projections canoniques $\pi_t(w) := w(t)$ $(w\in D,\ t\in\mathbb{R}_+)$ et la filtration-rendue continue à droite-qu'elles engendrent; c'est-à-dire $\mathbb{D} := (\mathcal{D}_t;\ t\in\mathbb{R}_+)$ avec $\mathcal{D}_t := \bigcap_{s\geq t} \sigma(\pi_u;\ u\leq s)$ $(t\in\mathbb{R}_+)$. Lorsque nous complétons les tribus par rapport à une probabilité P, nous écrivons $\mathcal{D}_t(P)$, $\mathbb{D}(P)$.

La loi du processus X, qui est la probabilité image de $\mathbb{P}$ par X, nous la notons $L(X)$. Si $(X, \mathbb{F})$ est une martingale, alors $(\pi, \mathbb{D})$ est une martingale sur l'espace probabilisé filtré $(D, \mathcal{D}, \mathbb{D}, L(X))$. Dans une telle propriété, ce qui est remarquable est le fait que, avec la probabilité $L(X)$ sur $(D, \mathcal{D})$, le processus (fixe) $(\pi, \mathbb{D})$ devient une martingale.

Nous dirons qu'une probabilité P sur $(D, \mathcal{D})$ est une <u>loi-martingale</u> (resp. loi-sous-martingale, resp. loi-surmartingale, resp. loi-semimartingale) si $(\pi, \mathbb{D})$ est une martingale (resp. sous-martingale, resp. surmartingale, resp. semimartingale) sous P. Appellons LMG l'ensemble de toutes les lois martingales sur $(D, \mathcal{D})$.

Sur LMG on peut considérer la topologie induite par la topologie étroite que est définie sur l'espace $MP(D, \mathcal{D})$ des mesures de probabilité sur $(D, \mathcal{D})$. Mais LMG n'est pas fermé pour cette topologie. L'un des premiers problèmes à aborder est celui de l'étude des sous-ensembles fermés de LMG (au sens de la topologie étroite). Voici un résultat partiel.

2. LEMME

*Les sous-ensembles* $U \subset LMG$ *qu'intègrent uniformément* $\pi_t$(*), *pour tout* $t \in \mathbb{R}_+$, *sont fermés pour la topologie étroite.*

Pour la démonstration voir REBOLLEDO [1].

3.

Un sous-ensemble particulièrement important de LMG est celui des élément $P \in$ LMG tels que $P(C) = 1$ où C est le sous-ensemble de D constitué par les fonctions continues de $\mathbb{R}_+$ dans $\mathbb{R}$ : nous l'appellons LMGC. Ainsi la mesure de Wiener que l'on construit sur D, est un élément de LMGC.

Les probabilités gaussiennes de LMGC sont caractérisées par la propriété suivante. Soit A: $\mathbb{R}_+ \to \mathbb{R}_+$ une fonction continue, croissante, nulle en 0, alors nous avons:

4. PROPOSITION

*Une probabilité* $P \in LMGC$ *est gaussienne de covariance* $K(s,t) = A(s \wedge t)$ $(s,t \in \mathbb{R}_+)$ *si et seulement si* $(\pi, \mathbb{D})$ *est une P-martingale continue de processus croissant associé A.*

C'est une modification d'un résultat classique caractérisant le mouvement Brownien dû à KUNITA et WATANABE [1].

Nous avons ainsi qu'à toute fonction continue croissante A, nulle en 0, nous pouvons lui associer une loi unique dans LMGC; nous désignons celle-ci par $P_A$ .

(*): c'est-à-dire, pour tout $t \in \mathbb{R}_+$, $\sup_{p \in U} \int_{\{|\pi_t| > c\}} |\pi_t| \, dP \xrightarrow[c \to \infty]{} 0$

Il existe donc un espace probabilisé filtré $(\Omega_A, F_A, \mathbb{F}_A, \mathbb{P}_A) \equiv (D, \mathcal{D}, \mathbb{D}, P_A)$ et une martingale gaussienne continue sur lui, $M_A \equiv \pi$, de processus croissant associé $\langle M_A, M_A\rangle = A$. Bien sûr, ni l'espace ni la martingale ne sont uniques: c'est seulement la loi de la martingale qui est unique.

5.

Retournons à l'espace $(\Omega, F, \mathbb{P})$ sur lequel nous considérons une filtration $\mathbb{F} = (F_t;\ t\in\mathbb{R}_+)$.

Soit M une martingale locale par rapport à $(\mathbb{F}, \mathbb{P})$. Nous introduisons la famille de processus

$$\alpha^\varepsilon[M](t) := \sum_{0\le s\le t} |\Delta M(s)|\ I_{\{|\Delta M(s)|>\varepsilon\}}, \qquad (5.1)$$

où $\varepsilon>0$, $t\in\mathbb{R}_+$ et la notation $\Delta M(s)$ est utilisée pour l'amplitude du saut en s : $\Delta M(s) = M(s) - M(s-)$.

Les processus croissant $\alpha^\varepsilon[M]$ sont localement intégrables, par conséquent, on peut trouver leur compensateur prévisible $\tilde{\alpha}^\varepsilon[M]$.

De même, les processus

$$A^\varepsilon[M](t) := \sum_{0<s\le t} \Delta M(s)\ I_{\{|\Delta M(s)|>\varepsilon\}} \qquad (\varepsilon>0,\quad t\in\mathbb{R}_+) \qquad (5.2)$$

sont à variations localement intégrables et leur compensateur prévisible $\tilde{A}^\varepsilon[M]$ existe.

Soient maintenant

$$\overline{M}^\varepsilon := A^\varepsilon[M] - \tilde{A}^\varepsilon[M] \quad \text{et} \quad \underline{M}^\varepsilon := M - \overline{M}^\varepsilon \qquad (\varepsilon>0). \qquad (5.3)$$

Le symbole $[\cdot,\cdot]$ est utilisé pour le processus "variation quadratique" . Soient alors.

$$\beta^{\varepsilon}[M] := [\overline{M}^{\varepsilon}, \overline{M}^{\varepsilon}]^{1/2} + [\overline{M}^{\varepsilon}, \underline{M}^{\varepsilon}]^{*} \qquad (\varepsilon>0)$$

(5.4)

(où $X^{*}_{(\cdot)} := \sup_{s\leq\cdot} |X(s)|$ pour tout processus X).

Ces processus sont croissants et localement intégrables, comme avant, nous écrivons les compensateurs prévisibles avec un "∿" : $\tilde{\beta}^{\varepsilon}[M]$.

Lorsque M est une martingale locale, localement de carré intégrable on peut considérer d'autres processus croissants, localement intégrables avec leur compensateur prévisible:

$$\sigma^{\varepsilon}[M](t) := \sum_{0<s\leq t} |\Delta M(s)|^2 I_{\{|\Delta M(s)|>\varepsilon\}} \qquad (\varepsilon>0,\ t\in\mathbb{R}_+)$$

(5.5)

$$\gamma^{\varepsilon}[M] := \langle\overline{M}^{\varepsilon}, \overline{M}^{\varepsilon}.\rangle + \langle M^{\varepsilon}, \underline{M}^{\varepsilon}\rangle^{*} \qquad (\varepsilon>0).$$

(5.6)

Les propriétés de tous ces processus peuvent être consultées dans REBOLLEDO [1] [2].

Ces processus nous serviront pour introduire les <u>conditions de raréfaction des sauts</u>: si $(M_n, \mathbb{F}_n)$ est une suite de martingales locales nous dirons qu'elle satisfait la <u>condition de raréfaction asymptotique des sauts, du 1^er^ type</u> (RAS (1)) (resp. la condition de raréfaction asymptotique forte des sauts du 1^er^ type RASF (1)) lorsque, pour tout $\varepsilon>0$, tout $t\in\mathbb{R}_+$, la suite $(\tilde{\beta}^{\varepsilon}[M_n](t);\ n\in\mathbb{N}$ (resp. la suite $(\tilde{\alpha}^{\varepsilon}[M_n](t);\ n\in\mathbb{N}))$ converge en probabilité vers 0.

Pour les martingales locales $(M_n, \mathbf{F}_n)$ que soient localement de carré intégrable, nous introduisons les conditions du second type: une telle suite satisfait la condition RAS (2) (resp. RASF (2)) si pour tout $\varepsilon>0$, tout $t\in\mathbb{R}_+$, la suite $(\tilde{\gamma}^{\varepsilon}[M_n](t);\ n\in\mathbb{N})$ (resp. $(\tilde{\sigma}^{\varepsilon}[M_n](t);\ n\in\mathbb{N})$) converge en probabilité vers 0.

Les relations entre ces conditions sont les suivantes: Pour les martingales locales, RASF (1) entraîne RAS (1); pour les martingales locales localement de carré intégrable, RASF (2) entraîne RAS (2) de même que RASF (1), par ailleurs RAS (2) implique RAS (1).

Il est important de noter que, dans ce dernier cas, si les martingales locales sont en outre quasi continues à gauche, alors RASF (2) et RAS (2) sont équivalentes.

Finalement, toujours dans le cas localement de carré intégrable, il faut remarquer que la condition RASF (2) est plus faible que celle que l'on connaît sous le nom de "condition de Lindeberg" (condition (L)):

$$\mathbb{E}\,(\sigma^{\varepsilon}[M_n](t)) \xrightarrow[n]{} 0 \qquad \text{(pour tout } t\in\mathbb{R}_+ \text{ tout } \varepsilon>0)$$

(Voir REBOLLEDO [2])

Nous verrons maintenant comment ces conditions interviennent dans les Théorèmes de la Limite Centrale.

## §2. LES THEOREMES DE LA LIMITE CENTRALE

Ce titre est un peu vague et il n'est adopté que pour des raisons historiques: le terme "limite centrale" évoque un certain parfum gaussien. Certains auteurs ajoutent le terme "fonctionnel" pour bien mettre l'accent sur la topologie faible utilisée (correspondante aux mesures définies sur un espace métrique de fonctions).

Précisons le type de propriété que nous voulons étudier.

1. DEFINITION

Une suite de probabilités $(P_n)$ sur $(D,\mathcal{D})$ satisfait la Propriété de la Limite Centrale (propriété LC) si elle converge étroitement vers une probabilité gaussienne centrée P de LMGC.

Comme nous l'avons signalé au paragraphe 1, n°4, toute probabilité gaussienne centrée de LMGC est associée de façon biunivoque à une fonction continue croissante A, nulle en 0. c'est pourquoi nous dirons que $(P_n)$ satisfait la Propriété de la Limite Centrale relative à A (propriété LC(A)) si $(P_n)$ converge étroitement vers $P_A$.

Etant donné une suite $(X_n)$ de processus càdlàg, nous dirons qu'elle satisfait la propriété LC (resp. LC(A)) lorsque la suite $(\mathcal{L}(X_n))$ des lois respectives satisfait LC (resp. LC(A)).

Lorsque A coïncide avec l'identité I sur $\mathbb{R}_+$, la mesure $P_I$ est la mesure de Wiener qui fait de $(\pi, \mathbb{D})$ un Mouvement Brownien. Reprenant les notations du début du paragraphe 1, le Théorème de Donsker s'exprime en disant que la suite $(M_n)$ satisfait la propriété LC(I).

Voici maintenant un résultat plus général.

2. THEOREME (De la Limite Centrale pour les Martingales Locales)

*Soit $((M_n, \mathbb{F}_n); n\in\mathbb{N})$ une suite de martingales locales, nulles en 0. Alors elle satisfait la propriété LC(A) dès que les deux conditions suivantes sont remplies:*

*RAS (1) et* (2.1)

*pour tout $t\in\mathbb{R}_+$, la suite $([M_n, M_n](t))$ converge en probabilité vers $A(t)$.* (2.2)

(REBOLLEDO [21])

Et lorsque les martingales locales sont localement de carré intégrable nous disposons de résultat suivant

3. THEOREME (De la Limite Centrale pour les Martingales Locales localement de carré intégrable).

*Soit* $((M_n, \mathbb{F}_n); n \in \mathbb{N})$ *une suite de martingales locales, localement de carré intégrable, nulles en 0.*

*Supposons que la suite satisfait RAS (2), alors les deux conditions suivantes sont équivalentes:*

$$[M_n, M_n](t) \xrightarrow[n]{\mathbb{P}} A(t) \text{ , pour tout } t \in \mathbb{R}_+ \text{ ; } \quad (3.1)$$

$$\langle M_n, M_n\rangle(t) \xrightarrow[n]{\mathbb{P}} A(t) \text{ , pour tout } t \in \mathbb{R}_+ \text{ . } \quad (3.2)$$

*Par ailleurs (toujours sous RAS (2)), si l'une des conditions équivalentes (3.1) ou (3.2) est satisfaite, alors* $(M_n, \mathbb{F}_n)$ *vérifie la propriété LC(A).*

(REBOLLEDO [1], [2], [3], [4], [10])

Ces résultats ont été déjà largement particularisés et appliqués à différents domaines des Probabilités.

Signalons, par exemple, l'étude asymptotique des chaînes de Harris réalisé par N.MAIGRET [1], puis par TOUATI [1]; les cas particuliers des Théorèmes de la Limite Centrale "discrets" obtenus par d'autres moyens par ROOTZEN [1] GANSSLER et HAUSSLER [1], applicables aux suites de variables aléatoires $(\xi_{nk})$ qui sont les "différences de martingales à temps discret"; l'étude de la convergence d'intégrales stochastiques dans les processus ponctuels vers une intégrale par rapport au mouvement Brownien et son application à la Statistique de Processus avec de données censurées (AALEN [1], [2]; GILL [1]; PONS [1]; REBOLLEDO [9]); l'analyse des sistèmes d'attente (REBOLLEDO [1] [11]; DUFLO [1]); l'analyse du comportement asymptotique des extrêmes dans GOUET [1], etc.

Ces différentes applications exigent une certaine familiarité avec le calcul de compensateurs prévisibles et de variations quadratiques. Nous n'en donneront pas de détails dans le présent résumé.

4.

Les théorèmes précédents motivent tout naturellement la question suivante: si $(M_n)$, suite de martingales locales, vérifie LC(A) , a-t-on la convergence en probabilité de $([M_n, M_n](t))$ vers A(t) pour tout $t \in \mathbb{R}_+$ ?

Il n'y a pas de réponse à cette question: elle est trop générale. Cependant, en ajoutant des conditions sur la taille des sauts on atteint une réponse affirmative.

La proposition suivante, obtenue en 1981 en forme indépendante par LIPTSER et SHIRYAEN [1], est un corollaire d'un théorème général portant sur les semi-martingales que j'ai publié en 1980 (REBOLLEDO [4] [5]).

5. PROPOSITION

*Soit* $((M_n, \mathbb{F}_n); n \in \mathbb{N})$ *une suite de martingales locales nulles en 0, satisfaisant l'hypothèse:*

*Pour tout* $t \in \mathbb{R}_+$, *la suite* $((\Delta M_n)^*(t); n \in \mathbb{N})$ *est uniformément intégrable.* (5.1)

*Alors les deux propositions suivantes sont équivalentes.*

$$[M_n, M_n](t) \xrightarrow[n]{\mathbb{P}} A(t) \quad \textit{pour tout } t \in \mathbb{R}_+ ; \qquad (5.2)$$

*La suite* $(M_n)$ *vérifie* LC(A). (5.3)

*Si en outre les martingales locales sont localement de carré intégrable, alors ces deux propositions sont équivalentes à*

$$\langle M_n, M_n \rangle(t) \xrightarrow[n]{\mathbb{P}} A(t) \quad \textit{pour tout } t \in \mathbb{R}_+ . \qquad (5.4)$$

6.

Nous reviendrons sur ce type de propriété en parlant des semimartingales. Cependant, je voudrais attirer l'attention du lecteur sur le fait que l'hypothèse (5.1) représente une condition plus forte que RAS (1) pour les martingales locales lorsque l'on suppose en outre (5.2) ou (5.3). Précisons:

*Si une suite de martingales locales vérifie* (5.1) *et* $(\Delta M_n)^*(t) \xrightarrow[n]{\mathbb{P}} 0$ *pour tout* $t \in \mathbb{R}_+$, *alors elle satisfait* RASF (1) (*donc, à fortiori,* RAS (1)).

Cela prouve que le Théorème de la Limite Centrale que l'on obtient en considérant comme hypothèse (5.1) et (5.2) est moins général que le théorème 2 ci-dessus. A ce propos il est intéresant de noter que le résultat fondamental de GANSSLER-HAUSLER [ 1] (à "temps discret") reprend les hypothèses (5.1) et (5.2) pour des suites de variables aléatoires constituant des "différences de martingales": soit $(\xi_{nk})$ une suite de variables aléatoires telles que $\mathbb{E}(\xi_{n,k} | F_{n,k-1}) = 0$ pour tous n,k , où $(F_{n,k};\ k \in \mathbb{N})$ est une filtration à temps discret; on pose $M_n(t) := \Sigma_{k=0}^{\tau_n(t)} \xi_{nk}$ , $F_{n,t} := F_{n,\tau_n(t)}$ où $\tau_n(t)$ es un changement de temps discret, i.e. $\tau_n$ est le compteur d'un processus ponctuel non explosif sur la droite réelle positive. Alors (5.1) et (5.2) se traduisent ainsi:

*Pour tout* $t \in \mathbb{R}_+$, $(\max_{k \leq \tau_n(t)} |\xi_{nk}|\ ;\ n \in \mathbb{N})$ *est une suite uniformément intégrable.* (6.1)

*Pour tout* $t \in \mathbb{R}_+$, $\Sigma_{k=0}^{\tau_n(t)} \xi_{nk}^2 \xrightarrow[n]{\mathbb{P}} A(t)$ (6.2)

Ces deux conditions entraînent RASF (1) dont la traduction au contexte "discret" donne:

*Pour tout* $t \in \mathbb{R}_+$, $\varepsilon > 0$, $\sum_{k=0}^{\tau_n(t)} \mathbb{E}(|\xi_{n,k}| I_{\{|\xi_{n,k}|>\varepsilon\}} | F_{n,k-1})$ *converge vers zéro en probabilité si* $n \uparrow \infty$. (6.3)

C'est-à-dire, (6.1) et (6.2) ⇒ (6.3) et (6.2) ⇒ LC(A).

7.

Passons maintenant aux semimartingales. Toute semimartingale $(X, \mathbb{F})$ nous allons la décomposer canoniquement comme

$$X = X(0) + M + B + \overline{X}^1 \qquad (7.1)$$

où :
- M est une martingale locale à sauts bornés par 1, donc, localement de carré intégrable;
- B est un processus à variations localement-intégrables et prévisible
- $\overline{X}^1(t) := \sum_{0<s\leq t} \Delta X(s) I_{\{|\Delta X(s)|>1\}} \qquad (t \in \mathbb{R}_+)$

Pour tout processus U à variations finies sur tout compact de $\mathbb{R}_+$, nous désignons par $V[U](t)$ sa variation totale $]0,t]$ $(t \in \mathbb{R}_+)$.

Pour tout $\varepsilon > 0$ nous définissons alors

$$\phi^{\varepsilon}[X](t) := \tilde{\alpha}^{\varepsilon}[M](t) + V[B](t) + V[\overline{X}^1](t) \qquad (t \in \mathbb{R}_+). \qquad (7.2)$$

8. THEOREME (de la Limite Centrale pour les Semimartingales)

*Soit* $((X_n, \mathbb{F}_n); n \in \mathbb{N})$ *une suite semimartingales nulles en 0 satisfaisant l'hypothèse:*

$$\phi^{\varepsilon}[X_n](t) \xrightarrow[n]{\mathbb{P}} 0 \quad \textit{pour tout } t \in \mathbb{R}_+, \textit{ tout } \varepsilon > 0 \qquad (8.1)$$

*Alors les deux propositions suivantes sont équivalentes:*

$$[X_n, X_n](t) \xrightarrow[n]{IP} A(t) \tag{8.2}$$

$$(X_n) \text{ satisfait la propriété } LC(A) \tag{8.3}$$

Ce résultat est un cas particulier du résultat fondamental de REBOLLEDO [4].

## §3. COMPACITE ETROITE RELATIVE ET TEMPS D'ARRET.

1.

L'étude de la compacité étroite sur MP(D, $\mathcal{D}$) peut être réalisée à l'aide de temps d'arrêt convenables.
la <u>ε-cribe fondamentable de temps</u> sur D:

Soit $\varepsilon>0$, nous définissons alors, pour tout $w\in D$

$$T_o^\varepsilon(w) := 0 \tag{1.1}$$

$$T_{n+1}^\varepsilon(w) := \inf \{t>T_n^\varepsilon(w): |w(t) - w(T_n^\varepsilon(w))|>\} \quad (n\in\mathbb{N})$$

Le résultat fondamental est alors le suivant

2. THEOREME

*Une famille $\Pi \subset MP(D, \mathcal{D})$ est relativement étroitement compacte si et seulement si toutes les conditions suivantes sont satisfaites:*

*Pour tout $N\in\mathbb{N}^*$*

$$\lim_{a\uparrow\infty} \sup_{P\in\Pi} P(\{w\in\mathcal{D} : |w(0)>a\}) = 0 \text{ et} \tag{2.1}$$

$$\lim_{a\uparrow\infty} \sup_{P\in\Pi} P(T_1^a \leq N) = 0$$

*Pour tous* $N \in \mathbb{N}^*$, $\varepsilon > 0$: (2.2)

$$\lim_{\substack{\delta \downarrow 0 \\ \delta < N}} \sup_{P \in \Pi} P(T_n^\varepsilon < N, \quad T_{n+1}^\varepsilon - T_{n-1}^\varepsilon < \delta) = 0,$$

*pour tout* $n \geq 1$.

*Pour tous* $N \in \mathbb{N}^*$, $\varepsilon > 0$: (2.3)

$$\lim_{\substack{\delta \downarrow 0 \\ \delta < N}} \sup_{P \in \Pi} P(T_n^\varepsilon < N, \quad T_{n+1}^\varepsilon - T_n^\varepsilon < \delta, \quad |\Delta w(T_{n+1}^\varepsilon)| > \varepsilon) = 0$$

*pour tout* $n \in \mathbb{N}$.

Voir REBOLLEDO [6], [7].

Ce Théorème se démontre par l'intermédiaire du Théorème de Prokhorov: on établit que la famille $\Pi$ est tendue (ou équitendue selon certains auteurs) et les temps de la cribe fondamentale servent à caractériser un ensemble relativement compact de D.

Etant donnée une suite de processus càdlàg $(X_n)$ nous dirons que la suite est tendue si $(L(X_n))$ est tendue (ou relativement étroitement compacte) dans MP(D, $\mathcal{D}$).

Le Théorème 2 sert à obtenir des conditions suffisantes pour qu'une suite de processus soit tendue. Dans cette direction, l'un des plus pratiques critères est le suivant, prouvé par ALDOUS [1] (dans mon article [1] j'avais démontré, de façon indépendante, un résultat voisin de celui de Aldous, c'est M.Métivier qui m'a fait connaître ce dernier travail qui a motivé mes recherches autour des conditions nécessaires et suffisantes pour avoir la compacité étroite de familles de mesures et qui ont abouti en Théorème 2)

3. THEOREME (Aldous)

*Soit* $((X_n, \mathbb{F}_n); n \in \mathbb{N})$ *une suite de processus càdlàg et supposons que les deux hypohèses suivantes soient vérifiées:*

*Pour tout* $N \in \mathbb{N}^*$, $((\Delta X_n)^*(N)\ ;\ n \in \mathbb{N})$ *et* (3.1)
$(X_n(0)\ ;\ n \in \mathbb{N})$ *sont* $\mathbb{R}$*-tendues ;*

*Pour tout* $N \in \mathbb{N}^*$, *pour toute suite* $(T_n)$ *telle que* $T_n$ *soit un* $\mathbb{F}_n$*-temps d'arrêt borné par* $N$, *pour toute suite réelle* $(\delta_n)$, $0<\delta_n<N$, *telle que* $\delta_n \downarrow 0$ *si* $n \uparrow \infty$ *on ait* (3.2)

$$X_n(T_n+\delta_n) - X_n(T_n) \xrightarrow[n]{\mathbb{P}} 0$$

*Alors la suite* $(X_n)$ *est tendue.*

Quand $(X_n)$ vérifie (3.2) on dit que la suite est <u>Asymptotiquement Uniformément Quasi-Continue à Gauche</u> (AUQG)

Lorsque les points d'accumulation des suites tendues sont portés par l'espace C des fonctions continues, nous parlons de <u>suites C-tendues</u>. Voici une caractérisation de telles suites:

4. COROLLAIRE

*Une suite* $((X_n, \mathbb{F}_n);\ n \in \mathbb{N})$ *de processus càdlàg est C-tendue si et seulement si les deux propriétés suivantes sont satisfaites.*

$(X_n(0);\ n \in \mathbb{N})$ *est* $\mathbb{R}$*-tendue et* $(\Delta X_n)^*(N) \xrightarrow[n]{\mathbb{P}} 0$ *pour tout* $N \in \mathbb{N}^*$. (4.1)

$(X_n)$ *est AUQG.* (4.2)

L'utilisation de ces théorèmes en Analyse Stochastique ne prend toute sa force qu'en étudiant des classes particulières de processus où l'on puisse disposer de caractéristiques "structurales" plus riches: par exemple, les semimartingales ou les martingales. L'idée maîtresse est de chercher à établir la tension de ces processus à partir de la tension de processus plus "simples": par exemple, les variations quadratiques ou les processus croissants associés.

Pour comparer la tension d'une suite de processus avec celle d'une autre, les notions de <u>domination de processus</u>

(LENGLART [1]) et de processus a accroisements dominés par un autre sont essentielles (REBOLLEDO [1], [7], [12]) Cela a donné lieu à quelques résultats d'ordre général que le lecteur intéressé pourra consulter dans mon article [7] et que nous ne reprenons pas à présent pour ne pas trop alourdir ce texte. Nous nous contenterons de montrer quelques critères de tension particuliers, relatifs aux martingales locales et aux semimartingales.

5. PROPOSITION

*Soit* $((M_n, \mathbb{F}_n);\ n\in\mathbb{N})$ *une suite de martingales locales, localement de carré intégrable. Supposons que la suite* $(\langle M_n, M_n\rangle)$ *vérifie le critère de Aldous (Théorème 3 ci-dessus); alors les trois suites* $(M_n)$, $(\langle M_n, M_n\rangle)$, $([M_n, M_n])$ *sont tendues.*

Pour la démonstration voir REBOLLEDO [1].

6. COROLLAIRE

*Etant donné une suite de martingales locales localement de carré intégrable* $(M_n, \mathbb{F}_n)$ *et si* $(\langle M_n, M_n\rangle)$ *est C-tendue, alors* $(M_n)$ *et* $([M_n, M_n])$ *sont tendues.*

*Si en outre,* $(M_n)$ *vérifie RAS (2), alors* $(M_n)$, $(\langle M_n, M_n\rangle)$ *et* $([M_n, M_n])$ *sont C-tendues dès que l'une quelconque de ces suites l'est.*

*En particulier, si les martingales locales* $M_n$ *sont continues,* $(M_n)$ *est (C-) tendue si et seulement si* $(\langle M_n, M_n\rangle)$ *est tendue.*

Pour les martingales locales (non localemente de carré intégrable) on a le critère suivant (voir mon article [7])

7. PROPOSITION

*Soit* $((M_n, \mathbb{F}_n);\ n\in\mathbb{N})$ *une suite de martingales locales, nulles en 0, telle que pour tout* $t\in\mathbb{R}_+$ $((\Delta M_n)^*(t);\ n\in\mathbb{N})$ *soit uniformément intégrable. Alors si* $([M_n, M_n])$ *est AUQG, les suites* $(M_n)$ *et* $([M_n, M_n])$ *sont tendues.*

8.

Finalement, montrons un exemple du type de critère que l'on peut obtenir avec des semimartingales. Pour cela considérons des semimartingales spéciales localement de carré intégrable, i.e. des semimartingales $X = X(0) + M + B$ où $M$ est une martingale locale localement de carré intégrable, $B$ est prévisible à variations localement intégrables.

Définissons

$$U[X] := \langle M, M\rangle + V[B] \tag{8.1}$$

Nous avons alors (REBOLLEDO [1])

9. PROPOSITION

*Soit une suite de semimartingales spéciales, localement de carré intégrable* $(X_n, \mathbb{F}_n)$. *Si* $(X_n(0))$ *est* $\mathbb{R}$ *- tendue et* $(U[X_n])$ *satisfait le critère de Aldous (en particulier si cette suite est C-tendue alors* $(X_n)$ *est tendue.*

Pour terminer, il faut signaler que les critères que nous venons d'exposer ont été le point de départ de plusieurs recherches sur la compacité étroite. Le dernier travail publié sur le sujet est la monographie de JACOD, MEMIN et METIVIER [1] où le lecteur purra trouver d'autres conditions suffisantes de tension.

## §4. PROBLEMES DE MARTINGALES

Nous traiterons maintenant un problème un peu plus général que celui de la propriété LC. Notre but est de construire quelques sous-ensembles remarquables de $MP(D, \mathcal{D})$.

Pour cela, nous allons nous restreindre à un cas très particulier de ce que l'on appelle les problèmes de semimartingales.

1.

Soient A et B deux processus càdlàg définis sur $(D, \mathcal{D})$, $\mathbb{D}$-prévisibles tels que A soit croissant et B, à variations finies sur tout compact de $\mathbb{R}_+$ ; tous les deux nuls en 0. Soit p une loi de probabilité sur $\mathbb{R}$.

Nous appellons PROB (p, A, B) l'ensemble de toutes les probabilités $P \in MP(D, \mathcal{D})$ pour les quelles $(\pi, \mathbb{D})$ devient une semimartingale spéciale localement de carré intégrable de décomposition $\pi = \pi(0) + M + B$ où $\langle M, M\rangle = A$ et p est la loi de $\pi(0)$ sous P. On dit que P est une solution du problème de semimartingales (p, A, B).

On appelle C l'ensemble de tous les couples (F, G) d'applications de D dans lui-même tels que l'application $w \to (w, F(w,\cdot), G(w;\cdot))$ de D dans $D(\mathbb{R}_+, \mathbb{R}^3)$ soit continue

*Dans ce qui suit, nous supposons que* $(A, B) \in C$. (1.1)

Considérons maintenant une suite $(X_n, \mathbb{F}_n)$ de semimartingales spéciales, localement de carré intégrable, décomposées canoniquement comme $X_n = X_n(0) + M_n + B_n$, $(n \in \mathbb{N})$.

La notation $A \circ X_n$ (ou $B \circ X_n$) désigne le processus défini comme $A \circ X_n(\omega, t) := A(X_n(\omega,\cdot), t)$ (resp. $B \circ X_n(\omega,t) := B(X_n(\omega,\cdot), t)$).

Nous conservons ces hypothèses et notations dans l'énoncé qui suit.

2. THEOREME

*Supposons qu'il existe une constante positive* $c>0$ *telle que*

$$\sum_{0<s\leq t} |\Delta X_n(s)|\, I_{\{|\Delta X_n(s)|>c\}} \xrightarrow[n]{\mathbb{P}} 0 \quad \textit{pour tout } t \in \mathbb{R}_+ \tag{2.1}$$

*Supposons en outre les hypothèses suivantes:*

*Il existe une fonction croissante positive* $\alpha$ *telle que* $A(w,t) \leq \alpha(t)$ *pour tout* $(w,t) \in D \times \mathbb{R}_+$ *;* (2.2)

*Les suites* $(A \circ X_n)$ *et* $(B \circ X_n)$ *sont AUQG;* (2.3)

*Pour tout* $t \in \mathbb{R}_+$, (2.4)

$$\sup_{s \leq t} |\langle M_n, M_n\rangle(s) - A \circ X_n(s)| +$$

$$+ \sup_{s \leq t} |B_n(s) - B \circ X_n(s)| \xrightarrow[n]{\mathbb{P}} 0$$

*La suite* $(L(X_n(0)))$ *converge étroitement vers* $p$.

*Alors la suite* $(L(X_n))$ *est tendue et tout point d'accumulation P de la suite est une solution du problème de semimartingales* $(p, A, B)$ *et en outre* $P((\Delta\pi)^* (\infty) \leq c) = 1$.

(2.5)

Pour la preuve voir REBOLLEDO [ 8].

On remarque que si (2.1) est vérifiée <u>pour tout c>0</u>, alors $P(C) = 1$ pour tout point d'accumulation de $(L(X_n))$, i.e. la suite est C-tendue.

3.

Pour illustrer ce Théorème étudions l'application aux diffusions. Soit $x \in \mathbb{R}$, considérons a,b deux fonctions réelles continues et bornées, a strictement positive.

Définissons

$$A(w,t) := \int_0^t a(w(s))ds; \; B(w,t) := \int_0^t b(w(s))ds, \quad (3.1)$$

$(w,t) \in D \times \mathbb{R}_+$ .

Selon STROOCK et VARADHAN, il existe <u>une seule solution P au problème de semimartingales $(\varepsilon_x, A, B)$ telle que P(C) = 1</u>: c'est la loi d'une diffusion partant de x.

Comme a et b sont bornées, existent $M_1, M_2 > 0$ telles que

$$|A(x,t)-A(w,s)| \leq M_1|t-s|; \quad |B(w,t)-B(w,s) \leq M_2|t-s|$$

cela entraîne immédiatement (2.3) et (2.2).

D'autre part, une application simple du Théorème de la Convergence Dominée de Lebesgue montre que $(A,B)\in C$. Par conséquent nous avons.

4. COROLLAIRE

*Pour que la suite $(X_n)$ converge en loi vers la diffusion définie par (3.1), partant de x, il suffit que pour tout $c>0$ (2.1) soit vérifiée et que en outre*

*Pour tout* $t \in \mathbb{R}_+$, (4.1)

$$\sup_{s\leq t} |\langle M_n, M_n\rangle(s) - \int_0^s a(X_n(u))du| + \sup_{s\leq t} | B_n(s) - \int_0^s b(X_n(u))du| \xrightarrow[n]{\mathbb{P}} 0$$

$$X_n(0) \xrightarrow[n]{\mathbb{P}} x. \quad (4.2)$$

C'est certainement dans la construction des solutions de problèmes de semimartingales où la convergence étroite prête et peut encore prêter une aide considérable. Nous n'avons fait qu' effleurer le sujet dans ce dernier paragraphe. Il y aurait encore beaucoup à dire sur les problèmes posés en termes des caractéristiques locales de semimartingales; les concepts de domaine d'attraction et de discrétisation de ces problèmes, etc. Nous arrêtons ici ce bref résumé, mais le chantier ne s'arrête point.

BIBLIOGRAPHIE

AALEN, O. [1] Statistical Inference for a Family of Counting Processes. Ph.D. Thesis, U. of California, Berkeley (1975).

—— [2] Weak Convergence of Stochastic Integrals Related to Counting Processes. Z.f.W <u>38</u> (1977). 261-277.

ALDOUS, D. [1] Stopping times and tightness. Ann Proba. <u>6</u> (1978), 335-340.

BILLINGSLEY, P. [1] <u>Convergence of Probability Measures</u>. John Wiley (1968).

DELLACHERIE - MEYER [1] <u>Probabilités et Potentiel I</u>. Hermann, Paris (1975).

———— [2] <u>Probabilités et Potentiel II</u>. Hermann, Paris (1979).

DUFLO, M. [1] Cours de 3è cycle. U. de Paris-Sud (1978).

GÄNSSLER - HÄUSLER [1] Remarks on the Functional Central Limit Theorem for Martingales. Z. für W. <u>50</u> (1979), 237-243.

GILL, R. [1] <u>Censoring and Stochastic Integrals</u>.

GOUET, R. [1] Thèse de 3è cycle. U. de Paris-Sud (1981).

JACOD, J. [1] <u>Calcul Stochastique et Problèmes de Martingales</u>. Lecture Notes in Maths. <u>714</u> , Springer (1979).

JACOD - MEMIN - METIVIER [1] On Tightness and Stopping Times. Stochastic Processes and their Appl. <u>14</u> (1983), 109-146.

LENGLART, E. [1] Relation de Domination entre deux processus. Ann. Inst. Herri Doincaré <u>13</u> (1977), 171-179.

LIPTSER - SHIRYAEV [1] On a problem of neccessary and sufficient conditions in Functional Central

Limit Theorem for Local Martingales. Z. für W. (1982).

MAIGRET, N. [1] Thèse de 3è cycle. Fac. de Sci. d'Orsay (1978).

PLATEN - REBOLLEDO [1] Weak convergence of semimartingales and Discretisation Methods. A paraître dans Stochastic Processes and their Applications.

PONS, O. [1] Thèse de 3è cycle. U. de Paris-Sud (1978).

REBOLLEDO, R. [1] La Méthode des Martingales Appliquée à l'étudé de la Convergence en Loi de Processus. Bull. Soc. Mathématique de France, Mémoire 62 (1979) 125 p.

——— [2] Central Limit Theorems for Local Martingales. Z. für W. 51 (1980), 269-286.

——— [3] Sur le Théorème de la Limite Centrale pour les Martingales Locales, C.R. Acad. Sci. Paris, Ser.A. t.288 (14 mai 1979), 879-882.

——— [4] Semimartingales et variations quadratiques. Conditions nécessaires et suffisantes de convergence en loi vers une martingale gaussienne. C.R. Acad. Sci. Paris, Ser.A, t.290 (5 mai 1980), 815-817.

——— [5] The Central Limit Theorem for Semi-Martingales: Necessary and Sufficient Conditions: Rapport U. de Nice (1980).

——— [6] Sur le Temps d'arrêt et la Topologie Etroite. C.R. Acad. Sci. Paris, Ser.A, t.289 (1979), 707-709.

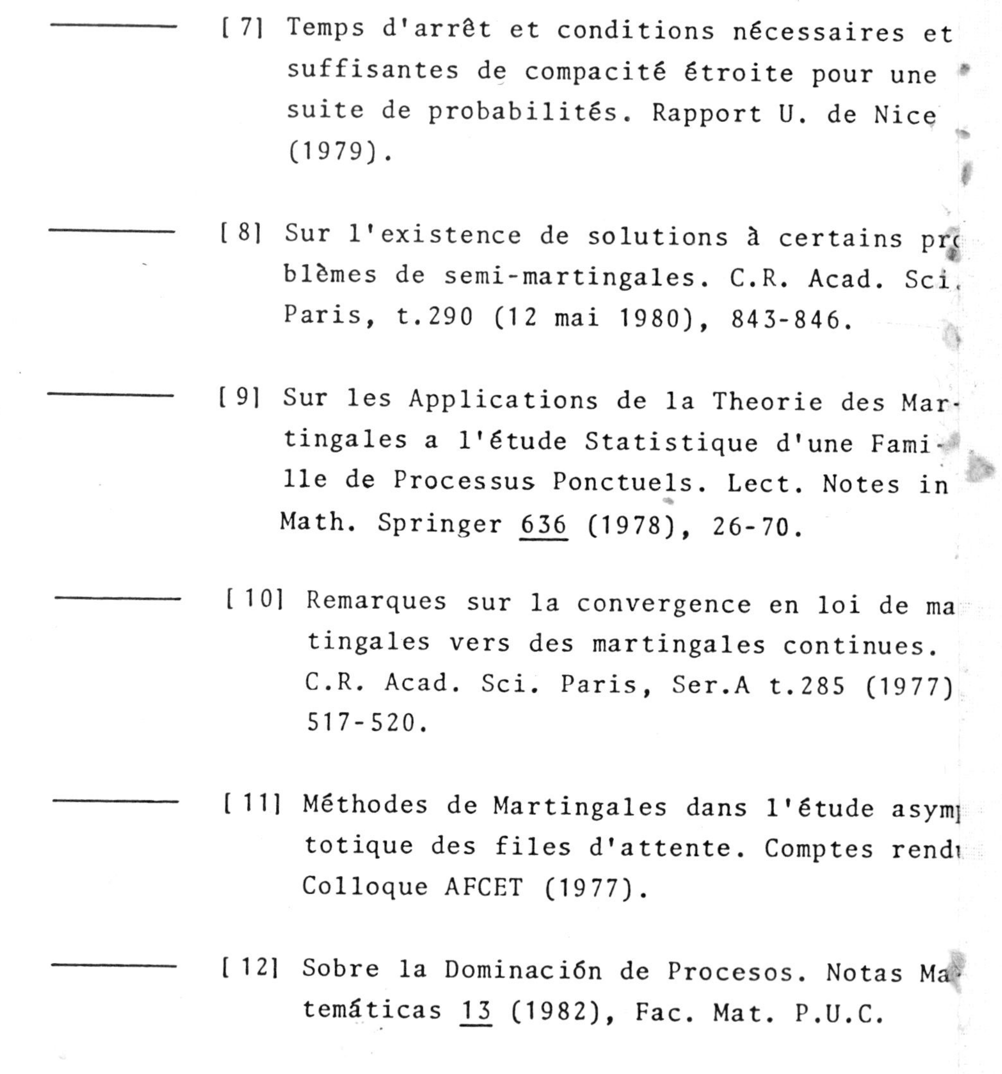

——— [7] Temps d'arrêt et conditions nécessaires et suffisantes de compacité étroite pour une suite de probabilités. Rapport U. de Nice (1979).

——— [8] Sur l'existence de solutions à certains pro blèmes de semi-martingales. C.R. Acad. Sci. Paris, t.290 (12 mai 1980), 843-846.

——— [9] Sur les Applications de la Theorie des Mar tingales a l'étude Statistique d'une Fami lle de Processus Ponctuels. Lect. Notes in Math. Springer 636 (1978), 26-70.

——— [10] Remarques sur la convergence en loi de ma tingales vers des martingales continues. C.R. Acad. Sci. Paris, Ser.A t.285 (1977) 517-520.

——— [11] Méthodes de Martingales dans l'étude asym totique des files d'attente. Comptes rendu Colloque AFCET (1977).

——— [12] Sobre la Dominación de Procesos. Notas Ma temáticas 13 (1982), Fac. Mat. P.U.C.

ROOTZEN, H. [1] On the Funcional Central Limit Theorem for Martingales. Z. für W. 38 (1977), 199-210.

STONE, C. [1] Weak convergence of Stochastic processes defined on semi-infinite time interval. Proc. A.M.S 14 (1963), 694-696.

TOUATI, A. [1] Thèse de 3è cycle U. de Paris-Nord (1980).